A Guide to Microsoft® Excel 2007 for Scientists and Engineers

09

Bernard V. Liengme
Microsoft Excel MVP

with

David J. Ellert
University of Southern Indiana
as co-author of Chapter 10

ELSEVIER

AMSTERDAM • BOSTON • HEIDELBERG • LONDON
NEW YORK • OXFORD • PARIS • SAN DIEGO
SAN FRANCISCO • SINGAPORE • SYDNEY • TOKYO
Academic Press is an imprint of Elsevier

Academic Press is an imprint of Elsevier
84 Theobald's Road, London WC1X 8RR, UK
30 Corporate Drive, Suite 400, Burlington, MA 01803, USA
525 B Street, Suite 1900, San Diego, California 92101-4495, USA

British Library Cataloguing-in-Publication Data
A catalogue record for this book is available from the British Library.

Library of Congress Cataloging-in-Publication Data
Application submitted.

ISBN: 978-0-12-374623-8

For information on all Academic Press publications,
visit our Web site at: http://www.books.elsevier.com

Printed and bound in the United Kingdom
Transferred to Digital Printing, 2010

Contents

Preface

This book is for people in technical fields, students and professionals alike. Its aim is to show the usefulness of Microsoft® Excel in solving a wide range of numerical problems. Excel does not compete with the major league symbolic mathematical environments such as Mathematica, Mathcad, Maple, and the like. Rather it complements them. Excel is more readily available and easier to learn.

The examples have been taken from a range of disciplines but require no specialized knowledge, so the reader is invited to try them all. Do not be put off by an exercise that is not in your area of interest. Each exercise is designed to introduce and explain an Excel feature. The two modeling chapters will help you learn how to develop worksheets for a variety of problems.

This is very much a practical book designed to show how to get results. The problem sets at the ends of the chapters are part of the learning process and should be attempted. Many of the questions are answered in the last chapter. The *Guide* is suitable for use as either a textbook in a course on scientific computer applications, a supplementary text in a numerical methods course, or a self-study book. Professionals may find Excel useful to solve one-off problems rather than writing and debugging a program, or for prototyping and debugging complex programs. A few topics are not covered by the *Guide,* such as database functions and making presentation worksheets. These are fully covered in Excel books targeted at the business community, and the techniques are applicable to any field.

I was agreeably surprised by the warm reception given the first and subsequent editions of the *Guide*. I am grateful for the many e-mailed comments and suggestions from readers and academics. The fourth edition has involved a major rewrite, not only because of how different Excel 2007 is from earlier versions but also to include more advanced material. I wish to thank David Ellert for his extensive input to the new chapter on VBA subroutines, John Quinn for his insightful comments on calculus and matrix algebra, and Robert van den Hoogen for kindly sharing his expertise in statistics. I am honored that Microsoft awarded me the Most Valuable Professional (MVP) in Excel both in 2007 and 2008 . My thanks are due to fellow MVPs for generously sharing their knowledge, in particular Jon Peltier and Bob Umlas. My final thanks go to my wife Pauline without whom this book would never have seen the light of day. However, I claim responsibility for all errors and typos.

I welcome e-mailed comments and corrections and will try to respond to them as soon as I can. Please check my web site and the *Guide*'s companion website www.elsevierdirect.com /companions/ 9780123746238 for supplementary material.

I hope you enjoy learning to "excel."

Bernard V. Liengme
bliengme@stfx.ca
http:/people.stfx.ca/bliengme

Conventions Used in this Book

Generally, in the chapters, the phrase *Excel 2007* is used to imply that a feature is new in this version or is very different from previous versions.

Information sidebars are used to give additional information, give reference, remind the reader of shortcuts, etc..

Information boxes in the left margin are used to convey additional information, tips, shortcuts, and the like.

Data that the user is expected to type is displayed in a distinctive font. This avoids the problems of using quotes. For example: In cell A1 enter the text Resistor Codes. Italics are used for new terms, to highlight Excel commands, for emphasis, and to avoid the confusion sometimes associated with quotation marks. Nonprinting keys are shown as graphics. For example, rather than asking the reader to press the Control and Home keys, we use text such as: Press Ctrl + Home. When two keys are shown separated by +, the user must hold down the first key while tapping the second.

In the Problems section of each chapter, an asterisk against a problem number indicates that a solution is given at the end of the book. Excel files for answered problems and additional files may be found on the companion website:

www.elsevierdirect.com/companions/9780123746238

Welcome to Microsoft Excel® 2007

The Excel Window

With Office 2007, Microsoft has abandoned the interface consisting of a menu and a collection of toolbars so common in all Windows applications until now. Their place has been taken by a ribbon divided into groups of commands located on named tabs. Figure 1.1 shows the Excel 2007 window; for this screen shot, the Excel window was "restored down" to occupy about half of the monitor screen.

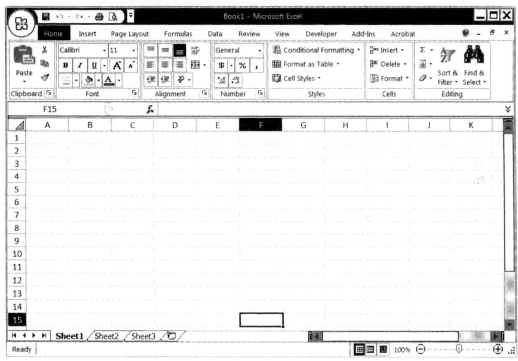

Figure 1.1

It is helpful to know the correct name for the various parts of the window. This makes using the Help facility more productive and aids in conversing with other users. It is recommended that you read this chapter while seated at the computer and experiment as you read it. Remember that pressing the Esc key will back you out of an action you do not wish to pursue.

Title bar: This is at the very top and displays the name of the currently opened file together with the phrase *Microsoft Excel*. To the right are the three controls to minimize, restore, and close the Excel application.

Office button: This is the name given to the colorful circle in the top left corner of the window. We can click on this icon to access commands relating to the file (open, close, and print). At the bottom of the Office dialog you will find a command to open a dialog to customize Excel. We will look at this in later chapters.

Quick Access Toolbar (QAT): This is the only toolbar in Excel 2007. When Excel 2007 is first installed, the QAT holds the commands Save, Undo, and Redo. However, it may be customized to hold others. Furthermore, one can change the location of the QAT from above the ribbon to below the ribbon.

Ribbon: The ribbon stretches across the window under the title bar. It holds every command that can be used within Excel 2007. In Figure 1.1 the Home tab has been selected, and the ribbon displays groups of commands that are accessed by clicking the appropriate icon. The Home tab holds mainly formatting commands. Use the mouse to open another tab by clicking it. We will see shortly that the ribbon can be minimized when you wish to see more of the document. The tabs shown in Figure 1.1 include Developer and Acrobat. We will learn in a later chapter how to add the Developer tab to the ribbon. The Acrobat tab gets added if you install Adobe® Acrobat®, which is not part of Microsoft Office products. Additional tabs (contextual tabs) get displayed when you are performing certain operations; so when you are working on a chart, the Chart tab appears.

The appearance of a tab will change with the amount of space allocated to the Excel window. Figure 1.2 shows the Home tab when Excel is in full-screen mode. Note how items that were arranged vertically in Figure 1.1 are now arranged horizontally.

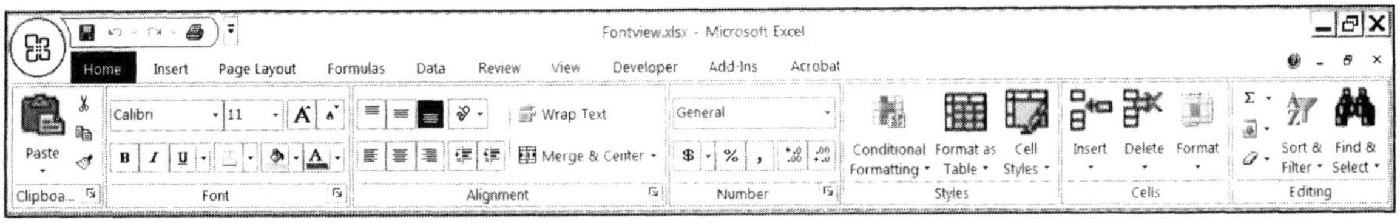

Figure 1.2

If you let the mouse pointer hover over a command icon, a screen tip will appear giving a brief description of the command's purpose.

Icons with solid inverted triangles ▼ (disclosure triangle) have associated drop-down menus that present further choices. When the diagonal arrow ◥ (the dialog launcher) on a group is clicked, a dialog box opens up. Generally these do not have new commands but present the group's commands in another way. Many dialog boxes have tabs either horizontally at the top or vertically at the left-hand side—see Figures 1.3 and 1.4. You can navigate from tab to tab with the mouse or with the arrow keys.

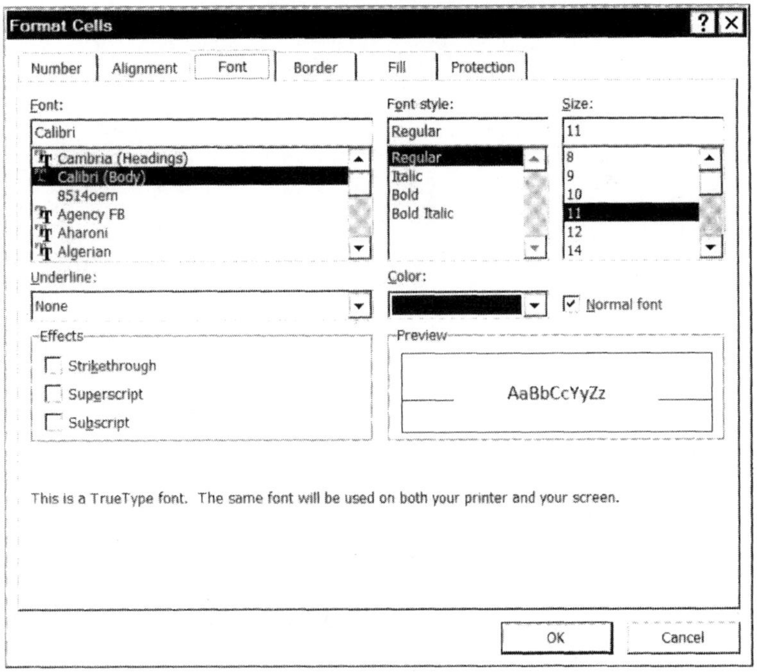

Figure 1.3

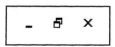

Help button: To the right of the tabs on the ribbon you will find the Help button. By default this connects you to the on-line help facility at the Microsoft Excel 2007 site.

Minimize, Restore, and Close buttons: To the right of the Help button are three tools used to minimize, restore, and close the worksheet. Note that we have one set of these buttons for the Excel application (on the title bar) and another (on the ribbon) for the current document.

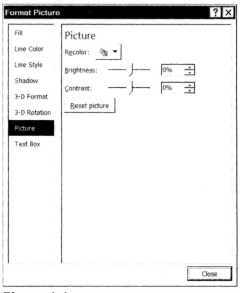

Figure 1.4

Formula bar and name box: Just under the ribbon is the formula bar with the name box to the left. In Figure 1.1 the name box is displaying F15. You will notice that both the F column heading and the 15 row heading are highlighted and that the cell at the intersection of this column and row is picked out by a border. We call this the *active cell*, and we say that the name box displays the reference (or address) of the active cell. Later we shall see that when the active cell contains a literal (text or number), the formula bar also displays the same thing, but when the cell holds a formula then the formula bar displays the actual formula while the cell generally displays the result of that formula.

Worksheet window: The worksheet window occupies most of the Excel space. In most cases this window displays a simple worksheet, but later we will see how to display two or more concurrently. A workbook may contain worksheets and chart sheets (collectively called sheets); we will concentrate on worksheets for now.

Sheet tabs: Below the worksheet window we have tools to navigate from sheet to sheet and to scroll a sheet horizontally. By default, Excel 2007 opens a new workbook with three worksheets that can be changed in the Options setting. To the right of the last sheet tab is a tool to insert a new worksheet. Let the mouse pointer hover over this tool to discover that the shortcut is ⌂ Shift + F11 . To the right of the sheet tabs is the horizontal scroll tool; the vertical scroll tool is on the right side of the worksheet.

Status bar: At the very bottom of the Excel window we have the status bar. To the left is the mode indicator. When you move to a cell this displays *Ready*; when you start typing it becomes *Enter*; if you double click a cell (or press the F2 key) it becomes *Edit*. We will ignore the second tool for now. To the right we have *Page View* buttons that let us display the worksheet in different ways, and the *Zoom* tool that enlarges/reduces the display. If we experiment with the *Page View* buttons, we may notice that the worksheet gets vertical and horizontal dotted lines. These show how much will fit on a printed page. Right clicking the status bar brings up a dialog box that allows you to customize the status bar.

Exercise 1: The Ribbon

The ribbon can be minimized so as to display about five more rows of a worksheet. Experiment with this as follows:

(a) Double click any one of the tabs on the ribbon. Most of the ribbon disappears leaving only the tabs—we say it is minimized.

(b) Click any one of the tabs once and the ribbon is maximized. It stays maximized until we activate a cell on the worksheet.

(c) Double click one of the tabs. The ribbon is permanently restored.

(d) Right click anywhere on the ribbon to bring up the shortcut menu. Click the Minimize Ribbon command.

(e) Restore the ribbon using the right click method.

Exercise 2: Quick Access Toolbar

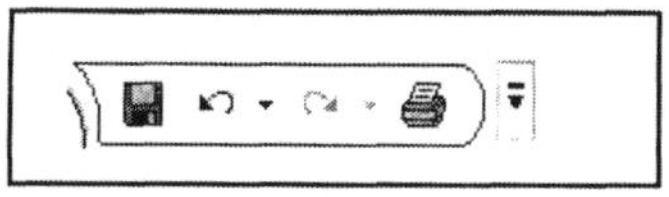

By default the QAT contains three commands: Save, Undo and Redo. We can add and remove commands in a number of ways; we look at one in this exercise. Do not overload the QAT; keeping it small so that all commands are easy to find preserves the intention expressed by *Quick* in its name. One command that may be handy to have on the QAT is the *Quick Print* command. This differs from the normal print command in that it is executed without first displaying a dialog box.

(a) Click the disclosure triangle ▼ to the right of the QAT to bring up the dialog shown in Figure 1.5.

(b) Click on the *Quick Print* item. The dialog closes, and QAT now displays a printer icon.

(c) Repeat the steps to add *Print Preview* to the QAT.

Next we will relocate the QAT. Right click on the QAT and in the shortcut menu select the item *Show Quick Access Toolbar Below Ribbon.*

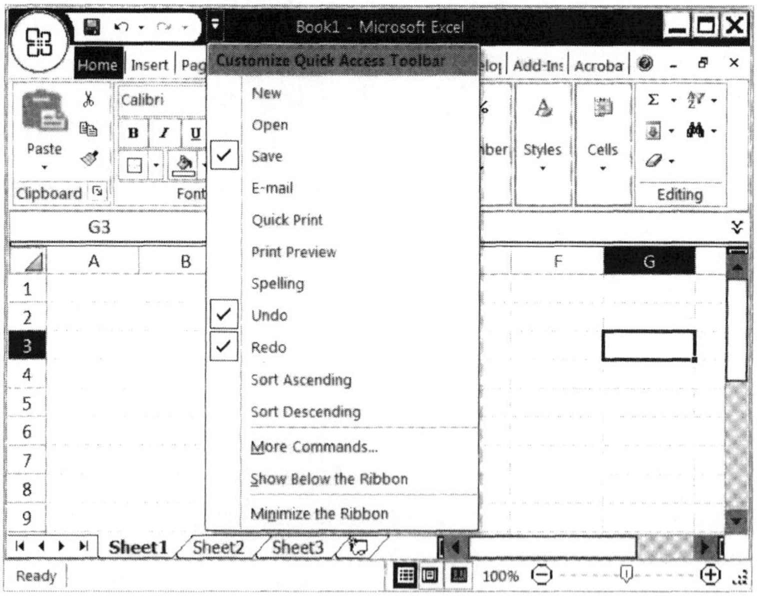

Figure 1.5

(d) We may restore the QAT to its original place in the same way or by using the shortcut menu we opened in step (a). Use either way to get the QAT above the ribbon.

Exercise 3: Working with Shortcuts

Some users prefer doing as much as possible from the keyboard rather than the mouse. Excel 2007 provides an extensive set of keyboard shortcuts. A full description of these would take many pages. So let's use the tools that Microsoft has provided—an on-line tutorial. This Exercise presumes you are connected to the Internet.

(a) Click the Help command. In the text box of the Help dialog type **shortcuts**. Either press the ⏎ or click the Search tool. When Excel responds with a list of topics, select Keyboard shortcuts in the 2007 Office system. Take some time running this very helpful tutorial; it will review many of the topics we have covered so far and then tell you all about keyboard shortcuts.

(b) When you return to Excel, you need to close the Help dialog by clicking its Close button ⊠ on the title bar.

The Worksheet

The worksheet window is the heart of the Excel application. It is here that we enter and work with data. It is helpful to learn some terms.

Columns and rows: A worksheet is divided vertically into columns and horizontally into rows. The intersection of a column and row forms a cell. At the top of the worksheet we have the column headers (the letters A, B, C...) and to the left the row headers (the numbers 1,2,3...). The last column is XFD (there are 16384 columns); the last row is numbered 1048576; thus a single sheet has some 17 billion cells. Your computer would need to have a very large amount of memory if you planned to fill every cell.

Cell: A cell is the unit on the worksheet; it may be empty or it may hold data. Generally cells are outlined by gridlines. However, it is possible to request Excel not to display gridlines for a particular worksheet. Note that gridlines are not printed unless otherwise specified in *Page Layout | Sheet Options*.

Active cell: If you click on a single cell on the worksheet, it is displayed with a solid border. We call this the *active cell*. The reference (such as A1) of the active cell is displayed in the name box. The correct term for the combination of column letter and row number (as in A1) is *reference,* but *address* is acceptable. What is not acceptable is *name* since this has a very special meaning in Excel. It is possible to configure Excel to use another reference system in which the top left cell is referred to as R1C1 but we shall not be concerned with that method. As noted above, the name box displays the reference of the active cell.

Data and Formulas: A cell may contain either data or a formula. Data and formulas are frequently entered by typing in the cell. How do we tell Excel we have completed your entry? There are a number of ways: pressing the Enter ⏎ key; pressing one of the arrow keys (⬇, ⬅, ➡, ⬆) or the tab key Tab⇆; or clicking the checkmark (✓) to the left of the formula bar. There is another method—clicking on another cell—but this is a very poor habit to pick up since the result when entering a formula is generally not what you want! The ⏎ key generally takes you down to the cell one below, but we can change this with an option setting to move one to the right. Data and formulas can also be placed in cells by copying (or cutting) them from other cells and then using the Paste command. The source cells can be in the same worksheet or in another worksheet, perhaps in another workbook.

Data: The data we entered into the cell can be one of four types. It could be text (such as the word Experiment), a number (123.45), a date (1/1/2008), or a Boolean constant (TRUE or FALSE).

Formulas: A formula always begins with an equal sign (=). It may contain only constants and cell references (=2*1.2345, =2*A2). It may also contain one or more functions (=SUM(A1:A10)). A formula normally displays a value in the cell; this can be any one of the data types listed above. So the cell containing the formula may display a value such as 6.28318, but when it is the active cell the formula bar may display the formula =2*PI(). If the formula fails, it may display (we say it returns) an error value. We start to use formulas in Chapter 2.

Formatting: This is the term used to describe changing how the value in a cell is displayed. We may format a cell to alter the font (typeface, size, color) and to add a border or a fill color. By far the most important aspect of this topic relates to numbers. In a newly opened worksheet every cell is formatted in what is called General. If I type 1.23456789 into a cell I will see 1.234567, and the formula =10*PI( ) will display 31.41953 since with the default column width a cell can display up to seven digits. We may widen the cell to display more digits. If I type 1234567890, Excel will widen the cell, but when more digits are used, as in 123456789012, Excel displays it in scientific notation as 1.234567E+11 (meaning 1.234567×10^{11}). Had the column been formatted to a narrow width beforehand, the result would show with fewer digits. We will see later that we may change the format of a number (for example, have =PI() display as 3.1). What is important to remember is that changing the format does not alter the actual stored value. We examine this in a later exercise, but it is good to learn early that there are stored values and displayed values.

Range: A range is a group of contiguous cells (see Figure 1.6). Note that technically a single cell is also a range: it is a range consisting of just one cell.

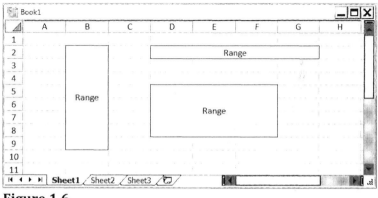

Figure 1.6

Excel 2007 Specifications

It is important to remember that Microsoft releases updates to all applications on a regular basis. Use the automatic update feature to stay current.

Excel Specifications: At some time you may need to know the answer to questions such as: What is the biggest number Excel can store? The information to answer this type of question is readily obtained from Help. Click the Help button and in the box type the word **specifications** (or just **specs**) and click the Search command. From the list of found topics select Excel Specifications and Limits. A screen shot showing part of the answer is shown in Figure 1.7.

Excel 2007 File Format: With Office 2007, Microsoft started using the Office Open XML format. This is not the place to delve into the technical aspects of this format. However, the reader should be aware that Office 2007 files are actually composed of a number of several XLM parts that are bundled into a zip-compressed file. This results in significant storage savings.

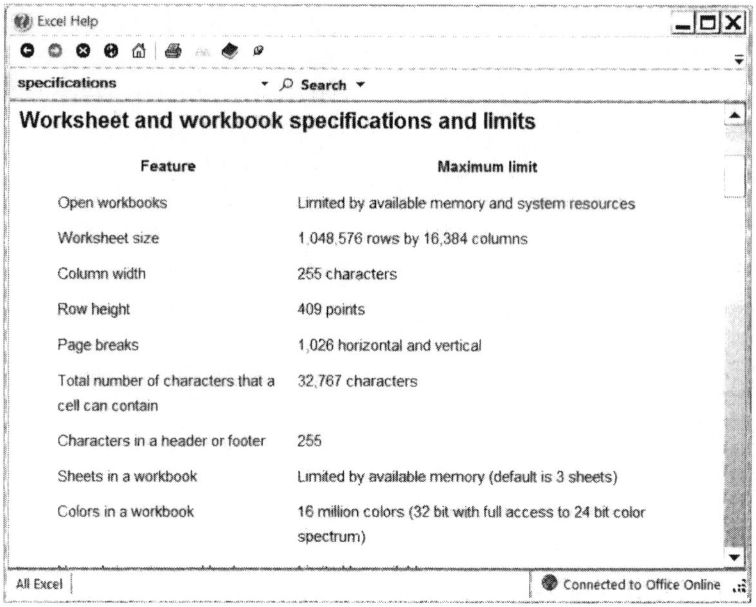

Worksheet and workbook specifications and limits

Feature	Maximum limit
Open workbooks	Limited by available memory and system resources
Worksheet size	1,048,576 rows by 16,384 columns
Column width	255 characters
Row height	409 points
Page breaks	1,026 horizontal and vertical
Total number of characters that a cell can contain	32,767 characters
Characters in a header or footer	255
Sheets in a workbook	Limited by available memory (default is 3 sheets)
Colors in a workbook	16 million colors (32 bit with full access to 24 bit color spectrum)

Figure 1.7

Excel 2007 files have one of these extensions: *XLSX*, *XLSM*, *XLTX*, *XLTM*, and *XLAM*. Until we begin to use VBA, all or our files will be saved as *XLSX* files. The letter *M* in a file extension denotes that it contains a macro, while *T* stands for template and *A* for add-in. We find out more as we progress through the book. We shall not be concerned with the binary format *XLSB*. The Microsoft website is the best source of information for the interested reader; search with the term Excel file formats.

One caveat: Some users have found that if they download an Office 2007 file (e.g., Book1.xlsx) from a website the download software mistakenly renames it Book1.ZIP, having detected its zip-compression attributes. All that is needed is to rename it back to the original extension before opening it with the Office 2007 application.

Compatibility with Earlier Excel Versions: There were various Excel file formats before Office 2007. However, Excel 97, 2000, 2002 (part of Office XP) and 2003 all had the same format and used the extension XLS for simple workbook files. While Excel 2007 can open files saved in the format of earlier versions, the converse is not true. When you open an XLS file in Excel 2007, the title bar will display the phrase Compatibility Mode and, unless you specify otherwise, the file will be save by Excel 2007 in the old XLS format.

To share newly created Excel 2007 files with users of say Excel 2003, you should save it in the XLS format; click the Office button, use the Save As command, and look for the Excel 97-2003 Worksheet. Should the workbook contain a feature not supported by Excel 97-2003 (for example, one of the functions new to Excel 2007), you will be given a warning. Also on the Office dialog under the Prepare tab there is a Compatibility tool that checks the workbook for Excel 2007 specific features that are not compatible with earlier versions. It is also possible for the other users to install the Microsoft compatibility utility that automatically converts Office 2007 files to the Office 97-2003 format. Search the Microsoft site using the term office 2007 compatibility.

Problems

If you are in a hurry, keep going to Chapter 2. If you like puzzle solving, try these problems. We will be covering the topics in subsequent chapters, but you may enjoy the challenge.

1. Type your name in any cell. Make it bold and italic. Find the commands to remove bold and italic. Hint: In the *Home*, look for an icon resembling an eraser.

2. In cell D1 enter this =TODAY( ) and press the ⏎ key. It should show the current date. Maybe it displays something like 15/3/2009; can you change it to 15-March-2009?

3. Copy the cell with your name. Paste it in another cell. Copy the cell with the date. Note the "ant track" running around the cell you copied. If you double click an empty cell, the track disappears and you can no longer paste. You have been using the Windows clipboard. Now click the *Clipboard* launcher on the Home tab (far left). This opens the Office Clipboard, which can hold more than one item. Experiment with it.

4. In A5 type the formula =22/7 and press ⏎. This gives an approximate value for π. Can you discover how to make this display with eight decimal places?

5. Type some numbers in the cells D1 to D5—later we will give this instruction as "put numbers in D1:D5." Click D6—or, in technical terms, make D6 the *active cell*. Look for the Σ icon (it is in *Home | Editing*). Click it to see what happens.

2

Basic Operations

This book is about problem solving so we shall spend little time on the preparation of presentation-worthy worksheets. We will give some information on how to make a worksheet more readable, but the emphasis is on mathematical operations in this chapter whose topics include:

- Entering numbers, including fractions and percentages
- Simple formulas such as =A1+B1+C1
- Range finders (colored borders showing what cells are used in a formula)
- Arithmetic operators +, −, * , / and ^
- The Evaluate Formula tool
- Error values such a #DIV/0! and #VALUE!
- Copying with command and shortcuts
- Formatting numbers
- The difference between stored and displayed values
- Round off errors resulting from the IEEE 754 standard.

If you are familiar with an earlier version of Excel, you may be tempted to skip this chapter. You are urged to at least read the Exercises to find out about new Excel 2007 features.

Exercise 1: Simple Arithmetic

Imagine that from time to time you are given some data consisting of rows of three numbers and you are asked to find the sum and product of the each triple (see Figure 2.1). Of course, this could be done with a simple calculator, but a spreadsheet offers three advantages: we can reuse our spreadsheet from day to day, we can see the values we have entered, and we can make a neat printout of the results. Our completed spreadsheet will look like Figure 2.2.

a	b	c	sum	product
1	3	4	8	12
4	5	6		
5	7	9		
6	8	3		

Figure 2.1

Chap2.xlsm

	A	B	C	D	E
1	a	b	c	sum	product
2	1	3	4	8	12
3	4	5	6	15	120
4	5	7	9	21	315
5	6	8	3	17	144

Figure 2.2

It is possible to have the worksheet display formulas in the cells using *Formulas / Formula Auditing / Show Formulas* (this is the top right icon) or the shortcut Ctrl + ` (the key to the left of 1 in the top row). This can be useful for printing documentation.

(a) In cells A1 to E1, enter the text shown in Figure 2.2. In A2:C3, enter the numbers shown. You will note that the text becomes left aligned in a cell while numbers are right aligned.

(b) Use the mouse to select A1:E1. On the *Home* tab, click the right alignment command in the *Alignment* group.

(c) Unless we have used a spreadsheet before we might be tempted to type =1+3+4. Try this (remember to press ↵ when finished) and it will give the correct answer, but this totally ignores the main idea behind a worksheet. We should not have to retype data. Wherever possible, formulas should refer to cell values.

(d) Click on D2 and tap the Delete key to remove this formula.

(e) Now type =A2+B2+C2 and click the check mark to the left of the formula bar when the formula is complete. Notice that, as you type, the status bar displays *Enter,* but once the checkmark tool (or ↵) is used, it shows *Ready*.

(f) Next we see another way to build a formula. In D3 type an equal sign (=), but rather than typing A3, click the A3 cell. Now type + and continue building the formula with this pointing method. Again click the checkmark to commit the formula once it is finished. In this case, pointing has little advantage over typing, but in other cases it has some advantages. It helps to ensure we reference the correct cell in a complex worksheet, and it is very useful to reference a cell on another sheet that could be in another workbook.

Of course, we do not have to rebuild the formula for every cell. We can copy from one cell to another. Here we look at two ways of doing this, and a little later we will see a third (and the fastest) method.

The first method uses the Copy and Paste commands located on the Clipboard group of the Home tab (far left)—see Figure 2.3.

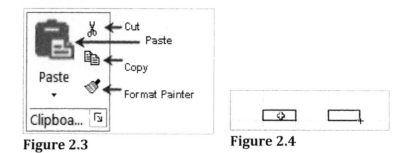

Figure 2.3   **Figure 2.4**

(g) With D3 as the active cell, click the Copy command. Select D4:D5 and use the Paste command (this is the larger icon on the group). Note how the cell we copied (D3) has a mobile dotted border. While this "ant track" is visible, the contents of the cell are still on the Clipboard and may be copied to any other range or cell. The ant track disappears as soon as you start to edit any cell but can also be removed by pressing Esc.

If you allow the mouse pointer to hover over the commands in the Clipboard, group screen tips pop up to tell the purpose and the shortcut keystrokes for each command. So Copy is Ctrl+C and Paste is Ctrl+V. Note that Ctrl+P is reserved for printing in most Windows applications.

(h) Delete D3:D5 and use the Ctrl+C and Ctrl+V shortcut method to fill them in.

(i) Delete D3:D5 again in preparation for another way to copy D2 down to D5. Move to D2 and note that the active cell border has a small solid square in the lower right corner; this is the *fill handle*. Carefully move the mouse pointer until it is over the fill handle—the pointer changes from an open cross to a solid cross (see Figure 2.4). Hold down the left mouse button and drag the solid cross down to D5. In step (l) below we shall see yet another method of filling a range.

For the final stage in this Exercise, we look at another approach to building formulas. Rather than type the formula in the cell, we will type it in the formula bar. There is an advantage to doing this when the formula is long, but we shall do it here for demonstration purposes.

(j) Make E2 the active cell. In the formula bar type =. Now complete the formula to be =*A2*B2*C2* either by typing or by pointing. Commit the formula with either ⏎ or the checkmark on the formula bar. Note that the multiplication operator is an asterisk (*).

(k) Lastly, we will fill in cells E3:E5. With E2 as the active cell move the mouse pointer over the fill handle (watch for the change from open to solid cross) and double click the fill handle. The formula from E2 is copied down to E5. This AutoFill feature can be used with vertical tables (data arranged in columns) but not with horizontal tables. It can be used to renew formulas when you change the top cell.

(l) Double click on any cell in the range D2:E5. Note the status bar displays *Edit*. But more importantly, observe the colored borders around the cells in the corresponding cells in columns A, B, and C. Excel uses these range finders to pictorially show you which cells a formula refers to.

(m) In the Quick Access Toolbar, click the Save command (picture of a floppy disc) and save the file as Chap2.xlsx.

Exercise 2: The Arithmetic Operators

The following table lists the arithmetic operators, their symbols, examples and order of precedence.

Operation	Symbol	Examples	Order of precedence
Negation	-	−1 and −A1	1
Percentage	%	5%	2
Exponentiation	^	2^3, A1^3,	3
Multiplication	*	2*3, 2*A1,	4
Division	/	2/3, A1/3,	
Addition	+	2+3, A1+3,	5
Subtraction	−	3−2, A1 −	

You may not be accustomed to treating the % symbol as an operator; essentially, it means divide the preceding number by 100. Exponentiation, of course, means raising a number to a certain power.

Excel evaluates formulas left to right using the order of precedence. So =40 – 20/5 will yield 36, not 10, since the division precedes the subtraction. We can override the order of precedence by the use of parentheses. Thus, =(40 - 20)/2 gives 10 because everything within parentheses is done first.

◢	A	B	C
1	2	-3	=A1+B1
2	1	2	=2*A2-B2
3	-1	2	=A1+A2+A3/B3
4	3	5	=(A1+A2+A3+A4)/B4
5	5	2	=-A5^B5
6	-5	2	=-A6^B6
7	5	5	=A7/(A7-B7)
8	5 apple		=A8+B8

Figure 2.5

Open the file made in Exercise 1 by clicking the Office button and locating Chap2.xlsx either in the Recent Document list or by using the Open command to browse through your Document folder.

(a) Click the Sheet2 tab (just above the status bar) to start a new worksheet or, if there is no Sheet2 tab, use the Insert Worksheet tool, which is to the right of the last tab in the sheet tab list.

(b) Look at Figure 2.5. Mentally compute each result.

(c) Create a worksheet using the values and formulas shown in Figure 2.5. Did you get the correct values? How about C5 and C6? Were there any surprises other than perhaps C7 and C8 which will be discussed shortly?

 Evaluate Formula

(d) To see how Excel performs a calculation, select each cell in turn and on the Formula tab use the Evaluate Formula tool in the Formula Auditing group. As you press the Evaluate button on the dialog, the formula is evaluated step-by-step. This tool is very useful with complex formulas; when you get an unexpected result, it helps you debug your work. You will note that -5^2 is evaluated as (-5)^2 since negation precedes exponentiation.

(e) Save the workbook.

We have not explained the results #DIV/0! in C7 and #VALUE! In C8. These are examples of error values. In the first evaluation step of the formula in C7 we get =5/0. Division by zero is said by mathematicians to be "undefined" so Excel tells us this is an error. A green triangle in the upper-left corner of a cell indicates an error in the formula in the cell. If you select the cell, the Trace Error button appears. Click the arrow next to the button for a list of options (refer to Figure 2.6). Experiment with this for yourself. You will find that Ignore Error removes the green triangle but this reappears if you double click the cell and press ⏎ without making changes to the formula.

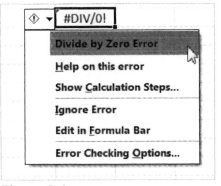

Figure 2.6

The #VALUE! error in C8 occurs because we are using the wrong data type: B8 contains a text value that is incompatible with the addition operator. We will soon discover that the SUM function ignores non numeric data, so it can be used in circumstances when we want to add values in a range, ignoring any text that happens to be in the range.

In addition to #DIV/0! and #VALUE!, other error values are #REF!, #NUM! , #NAME? and #N/A. We will discuss them as we proceed but for now note that each begins with a number (hash or pound) symbol. A worksheet displaying an error value has a mistake in it and needs attention except that #N/A is often acceptable and is taken to mean "not applicable" or "not available".

In Chapter 5 we meet some conditional functions that enable us to avoid error values such as #DIV/0! and #N/A in many circumstances.

Exercise 3: Formatting (Displayed and Stored Values)

This is a very simple Exercise, but it is most important for an Excel user to know the difference between a *displayed* and a *stored* value (Figure 2.7).

◢	A	B	C
1	Displayed and Stored Values		
2			
3	N	2+N	2*N
4	1.249	3.249	2.498
5	1.25	3.25	2.498

Figure 2.7

(a) Open your Chap2.xlsx workbook and move to Sheet3 or insert Sheet3 if necessary. Type the text shown in the rows 1 and 3 of Figure 2.7. Right align the entries in A3:C3 as we did in Exercise 1.

(b) In A4 and A5 type the value 1.249. With A5 as the active cell, locate the Decrease Decimal tool in the Number group of the Home tab (refer to Figure 2.8). Click this once to have A5 display 1.25.

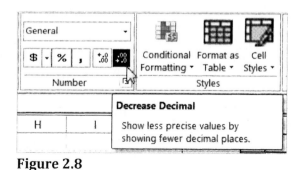

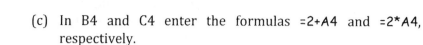

Figure 2.8

(c) In B4 and C4 enter the formulas =2+A4 and =2*A4, respectively.

(d) In B5 and C5 enter the formulas =2+A5 and =2*A5, respectively. For the purpose of this exercise, please do not copy them from the row above but type them in.

(e) Save the workbook.

Row 4 has no surprises, but look again at row 5. We know that A5 has the value 1.249 and that it was formatted to display only two decimal places. The addition of 2 gives 3.249, while multiplication gives 2.498. C5 displays the expected value, but B5 has a rounded result. It is a feature of Excel that a cell with a simple formula (such as =A1 which has no arithmetic operator or =A1+2 and =B1–3 with just the addition/subtraction operators) inherits the format of the referenced cell. Had we copied the formula from B4 to B5, the cell would have displayed 3.249.

Note that if you move to A5 the formula bar displays the stored value of 1.249 rather than the formatted value of 1.25. If we had wanted the formula to treat 1.249 as 1.25, then we could have used the ROUND function as shown later. Alternatively, we can have Excel treat all numbers to have the precision of the displayed values. This can be helpful in some financial accounting work but can lead to some confusion in other cases, so we shall not pursue this feature.

Exercise 4: Working with Fractions

Most of us work with decimal numbers, but there are still occasions when we would like to do some arithmetic with fractions. In this exercise we shall learn how to enter a number like 2 ¼ and how to have a number such as 14.6667 displayed as 14 10/15.

(a) Enter the text shown in A1, A3, and A6 of Figure 2.9.

(b) Enter the numbers in A4:C4. The number in A4 is entered by typing the **4** followed by a space and then **7/8**. Note that the formula bar displays 4.875. The ¾ is entered as **0 3/4**; if you enter only 3/4, Excel will be overhelpful and think you mean a date (3 Apr of the current year). If C4 has previously held a fractional value before you enter the 6, then the value will be displayed with spaces following it.

◢	A	B	C	D	E	F
1	Working with Fractions					
2						
3	Adding numbers to give an answer to the nearest 1/2					
4	4 7/8	3/4	6		11 1/2	
5						
6	Displaying and adding numbers to give answer in fifteenths					
7	3 3/15	6 5/15	5 2/15		14 10/15	

Figure 2.9

(c) In E4 enter =A4+B4+C4 either by typing or using the pointing method mentioned earlier. The result will be displayed as 11 5/8.

(d) We were set the task of having the result displayed to the nearest ½, so we need to format E4. Use *Home | Number* launcher; (southeast pointing triangle next to the word Number); refer to Figure 2.10. This opens up the Formatting dialog (Figure 2.11) from where we may select *As Halves ½* in the Fraction category.

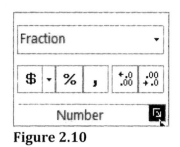

Figure 2.10

Figure 2.11

(e) Enter the values in A7:C7 and copy the formula E4 to E7.

(f) The result in E7 is actually 14.66667, but it displays as 14 ½ because the copy action copied both the formula and the format of E5. If you follow the instructions of step (d) above you will find that the only fractions Excel offers are halves, quarters, sixteenths, tenths, and hundredths. Are we out of luck in wanting fifteenths? No; all we need to do is move to the Custom category in the Format Cells dialog and in the Type box replace # ?/2 by # ?/15. Note there is a space after the # symbol.

(g) Save the workbook.

Exercise 5: A Practical Worksheet

In this Exercise we demonstrate a practical worksheet. An electrical engineer wishes to compute the effective resistance of four resistors in parallel; refer to Figure 2.12 for a diagram of what is meant by this and for the equation used to compute the answer. You are not expected to make the diagram! Also ignore the fact that gridlines are not seen and there are borders around some cells; we find how to do this shortly.

	A	B	C	D	E	F	G	H	I	J
1	Resistors in Parallel									
2										
3	Resistors	1240	1800	2000	4700					
4	1/R	0.00081	0.00056	0.0005	0.00021					
5										
6	1/R$_e$	0.00207		R$_e$	482					
7										
8	Alternative method									
9	R$_e$	482								
10										

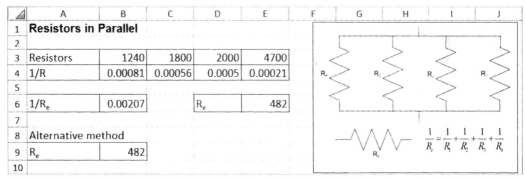

Figure 2.12

(a) Open Chap2.xlsx and use the *Insert Sheet* command (last item on the sheet tab list) to create Sheet4.

(b) Enter the text shown in A1, A3, A6, D6, A8, and A9. Enter the values shown in B3:E3.

(c) In B4 enter the formula =1/B3 and copy it across to E4 by dragging the fill handle.

(d) In B6 we compute 1/R1+1/R2+1/R3+1/R4 using the formula =B4+C4+D4+E4. You may wish to compose this using the pointing method. We will soon see how the SUM function can make life a bit simpler.

(e) In E6 we find the reciprocal of the sum of reciprocals with =1/B6 to give us Re.

In Chapter 5 we will redo this worksheet and use functions to get a better solution.

(f) In B9 enter the formula =1/(1/B3+1/C3+1/D3+1/E3) to demonstrate a shorter method.

(g) Use the Decrease Decimal tool on the *Home | Numbers* group to display E6 and B9 with no decimal places.

It is dangerous to rely on the results of any computer program (including an Excel worksheet), which has not been tested. Try your worksheet with some simple values such as four resistors of 2 ohms or four of 100 ohms. Does your worksheet agree with the results you computed in your head? This does not constitute a total validation of the worksheet, but it gives us more confidence in its results.

Does the worksheet have any limitations? Clearly it cannot be used for more than four resistors, but that is not a serious drawback from a practical point of view. How about fewer than four?

(h) Move to A7 and press the Delete key. Oh dear, our worksheet displays a number of #DIV/0! Error values. The blank value is treated as a zero value.

Let's think about the physical meaning of removing a resistor. It does not mean inserting a resistor of zero ohms; that would be a short circuit. Rather, it means replacing R4 by an infinite resistance. We cannot enter an infinity value, but we can enter a large number such as one mega ohm.

(i) Enter values of 2 for the first three resistors and 1E6 (you are familiar with this notation meaning 1×10^6 from your hand calculator) for the last one. Now compute the expected results in your head. Does the worksheet give a good answer? Of course, if the first three resistors have very big values, then our missing R will need to be very large, say 1E100. This is another problem we can solve more efficiently with functions.

Copying Formulas: What Happens to References?

We have seen in the last Exercise that when the formula =1/A4 was copied from B4 to B5, the formula was adjusted to =1/B5. This is very useful, but there are times when we want something else. First we need to understand how Excel goes about adjusting references when you copy a formula.

The formula in B4 was =1/A4; think of this as meaning =1/(the cell that is one column to the left, on the same row). This is what is meant by a *relative address*. The reference to A4 is interpreted relative to the cell that holds the formula. So when we copy this to

Note: a blank cell referred to in a mathematical formula is treated as having a zero value.

B5, it is still =1/(the cell that is one column to the left, on the same row), and this is, of course, represented by =1/A5. Now let's look at a problem where this automatic adjustment does not work for us. You may wish to make a worksheet of your own to experiment with this.

◢	A	B	C	D	E
1	Adjustment	1.10			
2					
3	Old table				
4	2	3	4	5	
5	5	6	7	9	
6					
7	New table				
8	2.2	3.3	4.4	5.5	
9	5.5	6.6	7.7	9.9	

Figure 2.13

In Figure 2.13 we have a range called *Old table* in A4:D5, and we wish to generate a range called *New table* in which the corresponding values in the old one have been adjusted by 10%. In cell B1 we have 1.10. In A8 we type =A4*B1 to get a value that is 10% higher than A4's value. All is well; we get 2.2 when A4 was 2. Now we use the fill handle to drag this down to A9, but we get =A5*B2. That is not what we wanted. We did want A4 to become A5 but wanted B1 to remain as B1. Furthermore, when we copy A8 to B8 we want =B4*B1, but we will get =B4*C2.

We solve this by using an *absolute reference* for B1. In A8 we type =A4*$B$1. When this is copied to A9 we get =A5*$B$1, and when it is copied to B* we get =B4*$B$1. You may think of a $ symbol (which has absolutely nothing to do with dollars, U.S. or otherwise) as an instruction to Excel to make no change to the column or to the row reference that follows it when the formula is copied. We say that a reference such as $B$1 is absolute since it is interpreted as *the cell in column B, row 1* regardless of what cell it appears in.

Now look at Figure 2.14 where we have started a multiplication table for a young person to test her math skills. Row 2 and column G have constant values (that is to say, not formulas). What formula shall we use in H3 such that we may copy it both across the row and down the column? We start with = G3*H2. When this is copied to the right, we want the first term (G3) still to point at G3, but when we go down a row we want it to be G4. So we see it's the G that is not to change. On the other hand, the second term (H2) must become I2 as we go across the row and H2 as we go down the column. So it is the 2 that should be immutable. This tells us to use =$G3*H$2. The term $G3 is interpreted as *the cell in column G, on the same row as the cell with the formula*. Part of the formula is absolute, part is relative. We call this a mixed reference; both $G3 and H$2 are mixed references.

G	H	I	J	K	L
Mulipication table					
	1	2	3	4	5
1	1	2	3	4	5
2	2	4	6	8	10
3	3	6	9	12	15
4	4	8	12	16	20
5	5	10	15	20	25

Figure 2.14

There is a keyboard shortcut method to add the $ symbols in absolute and mixed references. Let's say you have typed =G3. Now when you repeatedly press F4 this will cycle through =$G$3, =G$3, =$G3, and =G3. Suppose you have typed =G3*H2 and now wish to add the $ symbols. Position the mouse pointer at the beginning of, inside, or at the end of G3 and press F4.

What's in a Name?

There is an alternative to using an absolute reference. We can give a name to a cell. So in the worksheet shown in Figure 2.15 the user might have created a name for cell B1. Suppose the name *Adjustment* had been used (we will soon see how to do this), then the formula in A8 could have been =A4*Adjustment. Names are treated generally as absolute references; copying the formula from A8 to A9 would result in =A5*Adjustment. There are ways of making relative names, but we will not investigate that topic. You should also know that names are case insensitive; if the name was created as *Adjustment*, you can also use *ADJUSTMENT* or *adjustment* in a formula. Names may also be given to ranges. It should be obvious that we cannot assign any name that could be confused with a cell reference (such as X1); less obvious is that the names C and R are ruled out since these have special meaning for Excel. If you use a naming method that would result in an illegal name, Excel adds an underscore, as in X1_ and C_.

We will look at three ways of naming the cell B1 as *Adjustment*:

1. Using a neighboring cell as the source of the name. If the user

selects A1:B1 and then uses the command *Formulas | Defined Names | Create Names from Selection* (Figure 2.16, top left) a dialog (Figure 2.15, lower left) pops up. This permits us to specify which neighboring cell to use for the name. With A1:B1 selected one could use the shortcut $\boxed{\text{Ctrl}}$ + $\boxed{\text{⇧ Shift}}$ + $\boxed{\text{F3}}$ to open the *Create Names from Selection* dialog.

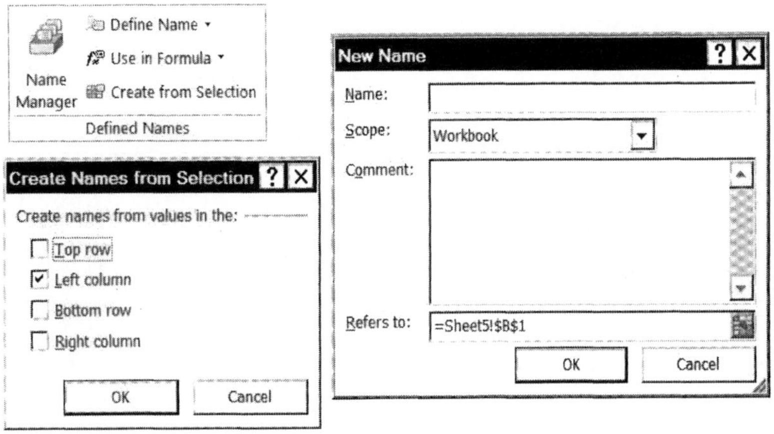

Figure 2.15

2. Using the command *Formulas | Defined Names | Defines Names* causes a dialog box (Figure 2.16, right) to pop up. This permits us to specify both the name and the cell(s) to which it refers. At this time we will not discuss the Scope option in the dialog.

3. With the cell selected, type a name in the name box and press $\boxed{\leftarrow}$ to complete it.

Method 3 is quickest for a single name; method 2 is very useful to name a ranges of cell that has text next to them which can be used for naming purposes; and method 1 is indispensable to give a name to a constant rather than to a cell (see below). We shall see later how to get a list of the names contained by a workbook.

Exercise 6: Another Practical Example

In this Exercise we construct a table showing giving the pressure of a gas at various temperatures and volumes using the van der Waals equation.

Figure 2.16 shows the completed worksheet. As we proceed we will learn how to spread an entry over a numbers of columns (Merge Cells) as is the title row 1, get superscripts as in CO_2, and use a custom format to display 250 as 250 K.

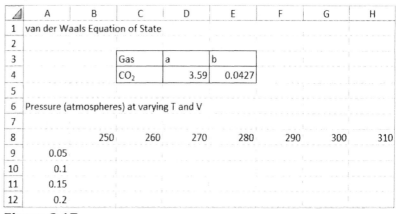

	A	B	C	D	E	F	G	H
1			The van der Waals Equation of State					
2								
3			Gas	a	b			
4			CO$_2$	3.59	0.0427			
5								
6			Pressure (atmospheres) at varying T and V					
7								
8			250 K	260 K	270 K	280 K	290 K	300 K	310 K
9	0.05 L	1374.21	1486.61	1599.02	1711.43	1823.84	1936.25	2048.65
10	0.10 L	-0.98	13.34	27.66	41.98	56.30	70.62	84.94
11	0.15 L	31.63	39.28	46.93	54.58	62.22	69.87	77.52
12	0.20 L	40.67	45.88	51.10	56.32	61.53	66.75	71.97
13	0.25 L	41.52	45.48	49.44	53.40	57.35	61.31	65.27
14	0.30 L	39.84	43.03	46.22	49.41	52.60	55.79	58.98
15	0.35 L	37.45	40.12	42.79	45.46	48.13	50.80	53.47
16	0.40 L	34.98	37.27	39.57	41.87	44.16	46.46	48.76
17	0.45 L	32.64	34.65	36.67	38.68	40.70	42.71	44.73
18	0.50 L	30.50	32.29	34.09	35.88	37.68	39.47	41.27

Figure 2.16

	A	B	C	D	E	F	G	H
1	van der Waals Equation of State							
2								
3			Gas	a	b			
4			CO$_2$	3.59	0.0427			
5								
6	Pressure (atmospheres) at varying T and V							
7								
8			250	260	270	280	290	300	310
9	0.05							
10	0.1							
11	0.15							
12	0.2							

Figure 2.17

Before we go too far, here are two reminders: (1) If you start something you do not wish to finish, such as typing in a cell that already has an entry that you would rather not destroy, then hit the (Esc) key; and (2) if you make a mistake such as deleting a complex formula and realize it in time, use (Ctrl)+Z to issue an Undo command.

(a)   Create a new worksheet in Chap2.xlsx. Start the worksheet by entering the values shown in Figure 2.17. Here is a quick way

Perhaps you want to increase the font size of the title in cell C1. Locate the two icons in Figure 2.8 with the letter A. These increase/decrease the active cell or current selection on one point each time they are clicked.

to get the numbers in row 8: Type 250 and 260 in B8 and C8; select the two cells; drag the fill handle to H8. Note the tooltip shows what number will appear in each cell as you move across. This feature is called *Auto Fill*. Use the same technique to fill A9:A18 with numbers from 0.05 to 0.5 in increments of 0.05.

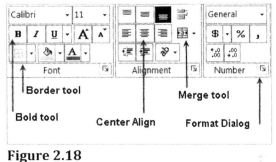

Figure 2.18

Now we will start our formatting using the tools shown in Figure 2.18.

(b) Select A1:H1 and click the Merge & Center tool to center the title over the first row. With the cells still selected, click the Bold tool to add emphasis. Now use the Merger & Center tool to get A6 centered over A6:H6

(c) Select C3:E4, click the downward pointing triangle next to the Border tool, and click on All Borders. With the range still selected, use the Center Align tool. Select B8:H18 and click the Border tool; note that the tool icon now shows four cells with borders around each after we used All Borders. Do the same with A9:A18.

(d) The numbers in row 8 are integers and need no special numeric formatting, but we wish to emphasize them as headers and show that these are temperature values in Kelvin. Select B8:H8 and click the Italic tool next to the Bold tool. With the cells still selected, launch the Format dialog and select Custom from the Category list on the left-hand side. Edit the Type box to read General " K" or 0 " K" with a single space before the K. In like manner italicize the numbers in A8:A18 and with the Increase Decimal tool get them to show two decimal places. Launch the Format Dialog (with the cells selected) and edit the Type box to read 0.00 " L" with a single space before the L. Some may argue that a lowercase l is the symbol for liter, but that is too easily confused for the digit 1.

(e) We wish to have *CO2* display as CO_2 with a subscript. Select C4 and in the formula bar use the mouse to select the 2 digit. Unfortunately, the common Office shortcuts ⌈Ctrl⌉+⌈+⌉and ⌈Ctrl⌉+⌈⇧ Shift⌉+⌈+⌉ do not work in Excel. Use *Home | Font | Font Launcher* and check the subscript box in the Format Dialog, (see Figure 2.19).

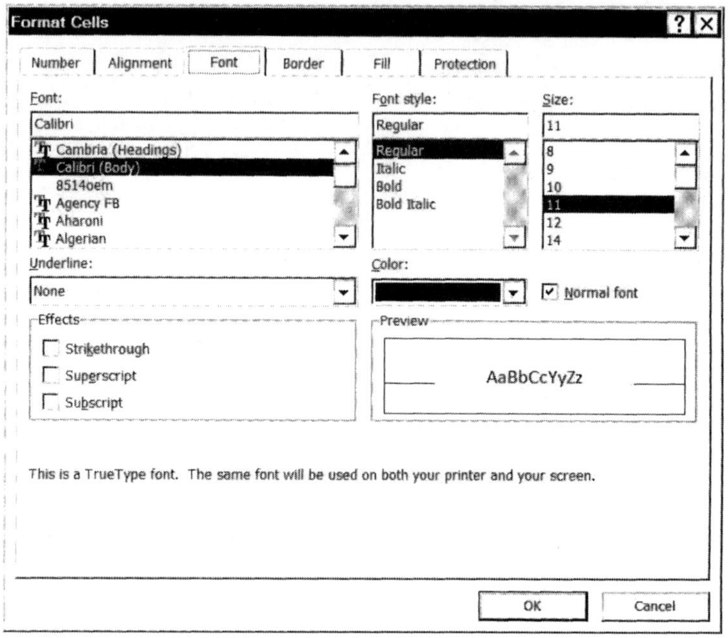

Figure 2.19

The next stage is to define some names. Our formula has three constants: a, b, and R. The first two vary from gas to gas, so we want to be able the change them on the worksheet, but the gas constant R is not something that can be altered (unless we want to work in other units) so we will "hide" it.

A name always refers to a *formula*. The formula may be: (i) a cell reference as in =Sheet1!$A$1; (ii) a range reference as in =Sheet1!A1:A10; (iii) a text or numeric literal as in =0.82058 or (iv) something more complex such as =SUM(Sheet1!$A$1:$A$8).

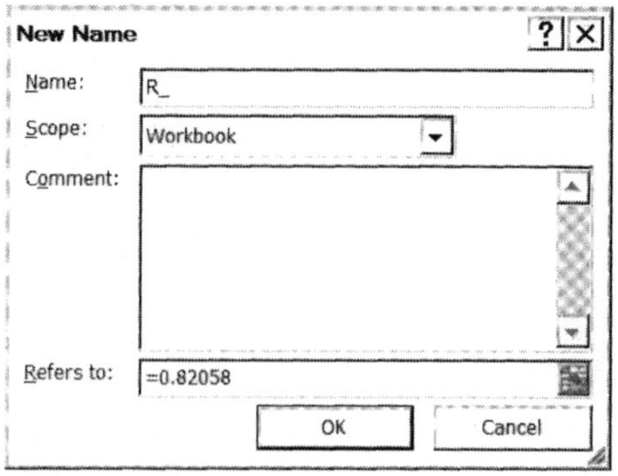

Figure 2.20

(f) Excel allows us to define a name that refers to a constant (numeric or textual). Use *Formulas | Defines Names | Define Name*. Complete the New Name dialog as shown in Figure 2.20. Note we must use R_ with an underscore.

(g) To have D4 and D5 named as a and b, respectively: select C4:D5, use *Formulas | Defines Names | Create* from Selection.

Finally, we need to enter a formula into B9 using relative, mixed, and absolute references in such a way that the formula may be copied across the row and down the column. From the van der Waals equation given at the start of the exercise our initial formula might be = (R_ * B8) / (A9 - b) - (a / (A9 * A9)). This has some redundant parentheses, but these were added to help read the formula. The named cells and the named constant will always be absolute references. We need the reference to B8 (the T term) to always point to row 8 and the reference to A9 (the V term) always to point to column A. This analysis of the problem helps us modify the formula to that shown below.

(h) In B9 enter the formula = (R_ * B$8) / ($A9 - b) - (a / ($A9 * $A9)). Using the fill handle, copy this across and down to fill B9:H18. Select a cell with that range and double click it. Observing the range finders, check that the formula does point to T and V correctly.

(i) Save the workbook.

Special Symbols, Subscripts and Superscripts

In the Exercise above we found how to convert CO2 to CO_2 with a subscript. It is obvious from Figure 2.20 that a similar method works to get superscripts. Superscripted digits 0, 1 , 2, 3 may be generated in another way since many fonts contain characters corresponding $^{\circ}$, ^{1}, ^{2}, and ^{3}. The default font for Excel 2007 is Calibri and this contains both sub- and superscripts for all the digits 0 through 9. There are no simple keys on the keyboard to get these, but in Microsoft Office we have two ways to produce both these and other symbols such as Δ, Σ, $\pm$, and ½.

The command *Insert | Text | Symbol* opens the Insert Symbol Dialog shown in Figure 2.21. The variety of available characters is greatly enhanced if the *Character Code From* box (in lower right corner) reads *Unicode*, rather than *ASCII*. It is unfortunate that subscripts 1, 2, and 3 are not in a range contiguous with the other superscript digits. Just select a character and click the Insert button to place one of these characters in a cell.

A second but more limited method to insert nonkeyboard characters consists of holding down the [Alt] key while typing a four-digit code beginning with zero on the numeric keypad. The requirement that the numeric keypad be used makes this method inconvenient for notebook users. The codes for some commonly used characters are shown in the top part of Figure 2.22. The symbols available this way and their codes are found from the Insert Symbol dialog with the *Character Code From* box reading *ACSII*.

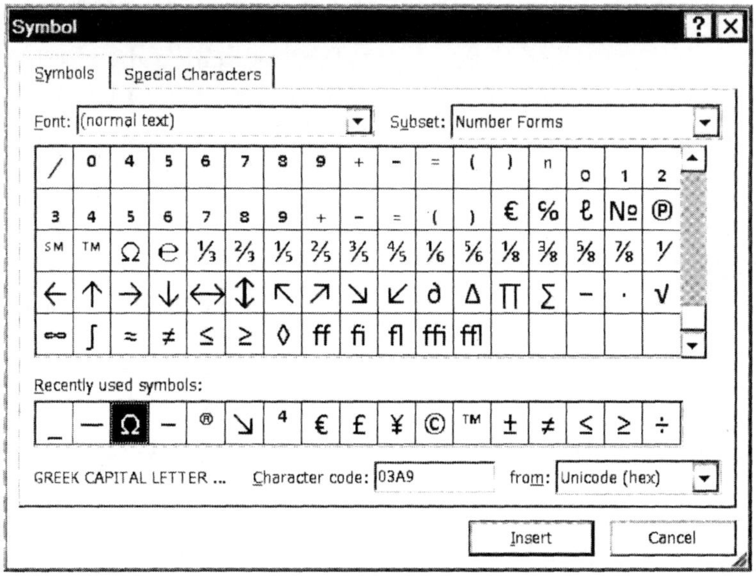

Figure 2.21

Special characters generated with [Alt]+0nnn on numeric keyboard

nnn	137	149	150	176	177	178	179	181	185	186	188	189	190	215	247
character	‰	•	—	°	±	2	3	μ	1	º	¼	½	¾	×	÷

Using the Symbol Font

Roman	a	b	c	d	e	f	g	h	i	j	k	l	m
Greek	α	β	χ	δ	ε	φ	γ	η	ι	φ	κ	λ	μ
Roman	n	o	p	q	r	s	t	u	v	w	x	y	z
Greek	ν	ο	π	θ	ρ	σ	τ	υ	ϖ	ω	ξ	ψ	ζ
Roman	A	B	C	D	E	F	G	H	I	J	K	L	M
Greek	Α	Β	Χ	Δ	Ε	Φ	Γ	Η	I	ϑ	Κ	Λ	M
Roman	N	O	P	Q	R	S	T	U	V	W	X	Y	Z
Greek	Ν	Ο	Π	Θ	Ρ	Σ	T	Y	ς	Ω	Ξ	Ψ	Z

Figure 2.22

Greek symbols may also be produced with the Symbol font. To produce ΔV, for example, type DV, select the D, and in the Format Cells dialog (Figure 2.19) select the Symbol font. Note that formatting is not visible in the formula bar, only in a cell. The correspondence between Roman and Greek letters is shown in the lower half of Figure 2.22.

The Microsoft Office shortcuts such as [Ctrl]+[+],for subscript, [Ctrl] +[⇧ Shift]+[+] for superscript, and [Ctrl]+[⇧ Shift]+Q for the Symbol font do not work in Excel; neither does the [Alt]+X method to change a Unicode to a character.

Mathematical Limitations of Excel

Like most computer programs, Excel uses the IEEE 754 standard for storing numbers. A number is converted from digital to a 64-bit binary representation. The fact that a finite number of binary digits are used has two major implications:

1. It limits the range of numbers that can be stored. Excel can store positive numbers from 1.79769313486232E308 to 2.2250738585072E-308. This limitation causes very few problems for users.

2. It limits the precision to which numbers can be stored. Excel has 15-digit precision. We need to understand the ramifications of the 15-digit precision limit.

Integer values: The integer value 123456789012345 with 15 digits is stored and displayed (when the cell has the appropriate format) exactly as typed in. An integer with more than 15 digits has trailing digits replaced by zero; it is not rounded. If we type1234567890123456, Excel stores and displays it as 1234567890123450; the trailing 6 becomes 0. There is no way to recover the lost precision. There is a simple solution when the "number" is actually just a string of digits as in a bank account number: precede the digits with a single quote (a.k.a., *apostrophe*), as in '1234567890123456. The single quote is not visible in the cell, nor will it appear in a printout but can be seen in the formula bar. Its purpose is to format the cell as text and a cell may contain up to 32,767 characters of text, but they may not all be displayed. Note that if 1234567890123450 is stored in A1 and '1234567890123456 is in A2, then =A1/2=A2/2 returns TRUE because when Excel does math on the text-formatted number, it must first lose the trailing 6 and put in a 0. However, we never actually do mathematical operations on bank account numbers or credit card numbers.

When the conversion from digital to binary leads to a result that is not exactly correct, we speak of *round-off errors*. Round-off errors are a fact of life in the computer world.

Real numbers: It often comes as a big surprise to many that Excel (and most other computer applications) cannot represent some real (that is, noninteger) digital numbers with total accuracy. But think about the fraction ⅓; we cannot display its value with complete accuracy in the decimal notation because it is 0.33333333…. and the threes go on forever. So it should not be too surprising that Excel cannot store with total accuracy the simple real number 0.1. In binary format this is 000110011001110011001111111….. and the ones go on forever.

The formula =(67.1-67.2)+1 is computed not as 0.9 but 0.899999999999991 because the intermediate calculation is 0.1, which gets stored with some inaccuracy. We have used 0.1 as an example; other decimal values can cause the same problem.

In A1:A6 enter these values 3.99, -25, 6.71, 6.59, 6.54, 1.17. In A7 write a formula to find their sum. Now use the Increase Decimals tool until you have 15 decimals showing. Surprise!

Moral: Unless you are working with integers, never test to see if one number is exactly equal to some other value. Never use =A20=B20 to see if the values computed by two methods give the same result. We can allow for round-off error by using formulas such as =ROUND(ABS(A20-B20), 10)=0 or =ABS(A20-B20)<1E-10 to see if the absolute difference in the two results expressed to 10 digits is 0 or not. The ROUND and ABS functions are explored in Chapter 4.

For more information on round-off errors, see any of the sites below, or search the Internet with the term Excel IEEE.

Floating-point arithmetic may give inaccurate results in Excel
http://support.microsoft.com/kb/78113/en-us.

(Complete) Tutorial to Understand IEEE Floating-Point Errors
http://support.microsoft.com/kb/42980.

What Every Computer Scientist Should Know about Floating Point
http://docs.sun.com/source/806-3568/ncg_goldberg.html.

Rounding Errors in Microsoft® Excel97
http://www.cpearson.com/excel/rounding.htm.

Visual Basic and Arithmetic Precision
http://support.microsoft.com/default.aspx?scid=http://suppor t.microsoft.com:80/support/kb/articles/Q279/7/55.ASP&NoW ebContent=1.

Play It Again, Sam

A very useful, but not well-known, Excel trick is the repeat shortcut. Here's how it works. Select a range of cells and add a border. Now select another range and press F4. That range gets the border. This trick works with many formatting features and can be a time saver.

Problems

1. (a) I typed 22.90 but Excel displayed 22.9. Name or describe the tool that I will use to see the trailing zero.

   (b) In cells A1:A2 I have typed 43.1, 43.2 and 1, respectively. In A4, I used the formula =A1-A2+A3 and it displays 0.9 as expected but if I display 15 digits I see 0.899999999999999. What do we call this type of error, and why does it occur?

   (c) I typed 1/1/2009 in a cell; find how to make the date display as 1-Jan-09. Experiment with Custom format to get 1-Jan-2009.

   (d) I typed 2/12 expecting to get a fraction, but Excel displayed 12-Feb (it might have shown 2-Dec had I been in Europe). What did I forget to do?

   (e) I wish to have column headers that read °F and ft³. How do I get this without formatting some characters as superscripts?

2. Referring to Figure 2.23, make a worksheet to compute the values in D using only cell references (no named cells).

⊿	A	B	C	D	E	F	
1	a	b	c				
2	5			25	$D2 : \sqrt{a}$	$D3 : \dfrac{c}{a^2 - b_2}$	
3	5	3	8	0.5			
4	729			9	$D4 : \sqrt[3]{a}$	$D5 : \dfrac{a - b}{b + c}$	
5	16	12	4	0.25			

Figure 2.23

3. Use a worksheet to answer these questions. (i) A basketball was found to have a volume of 440 in³. Does it conform to the NBA regulation that the circumference is to be 29.5 to 29.75 in. for male adults? (ii) A golf ball must not exceed 1.680 in. in diameter nor have a weight over 1.620 oz. What is the maximum density (oz/in³) of a golf ball? (iii) A soccer ball has a circumference of 28 in.; what is the area of the material required to make one? (iv) In SI units, water has a density of 1 gram / cm³. Given that 1 in. = 2.54 cm (this is the definition of the inch) and 1 oz = 28.3495231 grams, what is the density of water in lb/ft³?

The shapes are made with *Insert | Shapes.* The formulas could be typed into cells and the shapes given a transparent fill. Alternatively, the formulas could be added with *Edit Text* (right click a shape to see this) and the shapes given a pale fill.

4. *A contractor needs a worksheet to compute the number of packages of shingles to purchase for roofing jobs. Figure 2.24 shows a draft, but we need at least fives rows for each roof shape. The diagrams are optional. Draw up a list of possible improvements to this worksheet. As you learn more Excel, you might wish to return to this worksheet and make improvements.

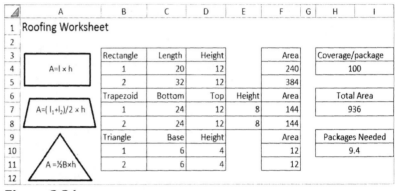

	A	B	C	D	E	F	G	H	I
1	Roofing Worksheet								
2									
3		Rectangle	Length	Height			Area		Coverage/package
4	A=l×h	1	20	12			240		100
5		2	32	12			384		
6		Trapezoid	Bottom	Top	Height		Area		Total Area
7	A=(l₁+l₂)/2×h	1	24	12	8		144		936
8		2	24	12	8		144		
9		Triangle	Base	Height			Area		Packages Needed
10		1	6	4			12		9.4
11	A=½B×h	2	6	4			12		
12									

Figure 2.24

5. *Columns A and B of Figure 2.25 show data collected by a team of students working on a solar car. The objective is to compute the average speed between each pair of data points. *Hint*: after typing the A5 value, use =A5 in A20, then format this as *General* to display the value 0.04167. How does this relate to the 1:00 in A5? Excel stores dates and times in day units! Column I gets the results in an alternative way not using the data in columns D and E. What formulas are used in D5, E5, G5, and I5?

	A	B	C	D	E	F	G	H	I
1	Average Speed								
2									
3	Clock time	Odometer Reading		Time interval	Distance		Average Speed		MPH
4	0	100.0							
5	1:00:00	158.7		1:00:00	58.7		58.70		58.70
6	2:04:23	218.4		1:04:23	59.7		55.64		55.64
7	2:56:24	267.5		0:52:01	49.1		56.64		56.64
8	3:45:23	315.8		0:48:59	48.3		59.16		59.16
9	4:12:00	340.3		0:26:37	24.5		55.23		55.23
10	5:34:03	422.4		1:22:03	82.1		60.04		60.04

Figure 2.25

6. The Antoine equation is as follows,

$$\log_{10}(p^*) = A - \frac{B}{T+C}$$

where p^* is in mmHg and T is the temperature in degrees Celsius. On a worksheet, the values of A, B, and C for a certain substance are stored in cells A3:C3, while A4 has a temperature value in Celsius. What Excel formula would you use in B4 to compute $\log_{10}(p^*)$? How would you modify this if A4:A20 had temperature values and you wished to copy the formula down to B20? What formula in C4 would compute p^*? Data to test your formula: for benzene A, B, and C are 6.90565, 1211.033, and 220.790, respectively, and benzene boils at 80.1°C.

7. If a volume V_1 of water at temperature T_1 is mixed with another volume V_2 of water at temperature T_2, the resulting temperature T_f can be found using $V_1(T_f-T_1)-V_2(T_f-T_2)=0$. Construct a worksheet similar to that in Figure 2.26 to give T_f.

◢	A	B	C	D	E	F
1	Final Temperature					
2						
3	First volume		Second volume			T_f
4	V_1 (gals)	T_1 (°F)	V_2 (gals)	T_2 (°F)		
5	25	50	60	180		141.76

Figure 2.26

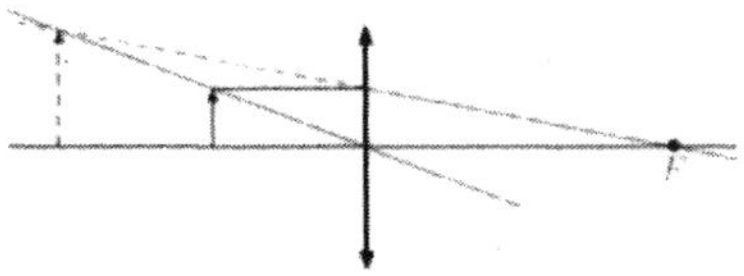

8. The *thin lens equation* shown in Figure 2.27 gives the relationship between the distance of the object (u) from the lens, the distance of the image (v) from the lens, and the focal length (f) of the lens. In the form shown we use the Cartesian convention: the incident light shines left to right, distances to the left of the lens are considered negative, and convex lenses have positive focal lengths, while concave lenses have negative f values. An object is placed 12 cm from a convex lens with focal length 18 cm. As the light comes from the left, the object must be placed to the left. Show that the image is at distance -36 cm. Since the object is on the same side as the image, it is imaginary. Construct a worksheet similar to Figure 2.27. Can you get the same result without using the "helper" column with the reciprocal values?

◢	A	B	C	D	E	F	G	H	I
1	Lens Equation								
2									
3	Focal length of lens		f	18	1/f	0.055556			
4	Object to lens		u	-12	1/u	-0.08333			
5	Image to lens		v	-36	1/v	-0.02778			

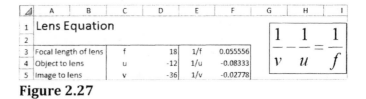

$$\frac{1}{v} - \frac{1}{u} = \frac{1}{f}$$

Figure 2.27

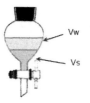

Vw

Vs

9. In the solvent extraction process, some solvent is mixed with an aqueous solution, shaken, and drained off taking some of the solute for later recovery. The process can be repeated with fresh aliquots of solvent to recover more solute. Let the volumes in each step be V_w and V_s for the water and the solvent, respectively; and let m_0 and m_1 be the mass of solute in the water before and after an extraction step. The *distribution coefficient* may be written as shown below. This will be constant for a given solvent-solute pair.

$$K_d = \frac{m_1 / V_w}{(m_0 - m_1) / V_s}$$

A 100 ml aqueous solution containing 5 g of solute is extracted four times with 75 ml of solvent. The value of K_d for this solvent-solute pair is 0.43. Construct a worksheet to: (i) Find the mass of solute remaining in the water after each of four consecutive exactions. (After the first extraction, m_1 = 1.8), and (ii) show that four extractions of 75 ml is more efficient than two at 150 ml.

10. You wish to restrict the values that may be typed into cell B2 to integers in the range 10 to 100, inclusive. Experiment with *Data | Data Tools | Validation* to get this condition. Write a paragraph telling others how to do this.

11. You wish to select all the cells in a worksheet so that you can change the font. Use Help to find two ways to do this. Write a short paragraph telling others how to do this. Remember to explain what is meant by *current range.*

12. In columns A1:B100 you have some numbers. In C1 you have the formula =A1/B1, and this is copied down to C100. Some B values are zero giving #DIV0! errors. Experiment with *Home | Editing | Find* such that you are able to select these cells and delete them. Write a paragraph telling others how to do this.

13. In A1 type the number 1. Select the cell and drag the fill handle down to A5. You get a series of 1's. Return to A1 and experiment with *Home | Edit | Fill* to make the series 1...10

   In B1 type a date such as 1/1/2009. Select the cell and drag the fill handle down to B5. You get a series of increasing dates. Excel is being helpful but this may be not what you want. Delete B2:B10. Hold Ctrl as you drag the fill handle of B1. Now the same date is repeated.

   What happens if you have a simple number in C1 and you hold Ctrl as you drag its fill handle?

14. Excel has many shortcuts. For example, type a number in D1 and move to D2. Now use Ctrl+' (the key next to ↵)and the same number appears. Move to an empty cell and use Ctrl+; and you get today's date. Search the Internet with the term **Excel shortcuts** to learn more about shortcuts. Don't try to learn them all—just the ones you might need in the near future. If you have a date such as 1/1/2009 in F1 and use the copy-cell-above shortcut (Ctrl+') what happens? You may get the same date but you are more likely to get a five-digit number such as 39814. What does that number mean?

3

Printing in Excel

Even in this so-called paperless-office world we still need to print our worksheets from time to time. Although we have hardly started on our study of Excel 2007, this is as good a place as any to discuss the printing process. In this chapter we shall learn about:

- The Print dialog and its options
- The Print Preview feature that can save paper wastage
- The Print-Area and how to set it
- Setting the margins and orientation
- Setting options such as printing gridlines and row/column headers
- Getting Excel to print the same rows and or columns on every page
- Printing a selection
- Inserting page breaks
- Changing the paper orientation

Excel will not respond to any print command if there is no printer installed on the computer. That does not mean an actual printer must be attached but rather a printer driver must be installed on the PC. You can perform most of the exercises in this chapter with just a printer driver installed. Furthermore, Excel will not print or show print preview if the worksheet is empty. This is true even if you have added a header or footer.

Exercise 1: Quick Print and Print Preview

As we shall soon see, Excel normally opens a dialog box when you issue a print command. This enables the user to make certain adjustments. However, if you just want to print the worksheet without making changes, that is also an option.

(a) Open Sheet6 of Chap2.xlsx where we did the van der Waals calculations.

(b) Print the worksheet using either *Quick Print* command on the *Quick Access Toolbar* (Exercise 1 of Chapter 1) or with the command *Office | Print | Quick Print*.

(c) Retrieve the printout from the printer.

(d) Now use the command *Office | Print | Print Preview* and compare your printed page with the screen—they should be identical. Note the *Close Print Preview* tool on the Print Preview tab; this returns you to Normal view (Figure 3.1)

Note that on this worksheet the last cell that has an entry is H18. That means that by default Excel will print the range A1:H18 on as many sheets of paper (pages) as necessary. With the font size we have used, our worksheet will now occupy less than one page.

Note: the shortcut for Print Preview is [Ctrl]+[F2]

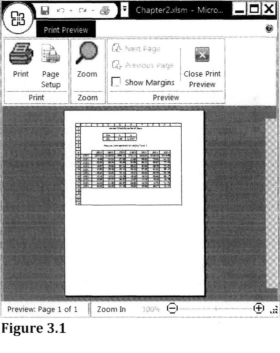

Figure 3.1

Figure 3.2

Once you have issued a print command (even Print Preview) Excel adds dotted lines to your worksheet showing the page breaks. The exact position is printer dependent. On the author's worksheet the first vertical dotted line was between columns I and J and the first horizontal one, between rows 46 and 47. These automatic page breaks, unlike manually entered page breaks, cannot be removed. If they clutter your worksheet, just close the file and then reopen it—remembering to save it first if needed.

The *Page Layout* tab contains the *Themes* group which we will not use; the *Page Setup* group with controls for margins, orientation (portrait and landscape), paper size and others; the *Scale to Fit* group and the *Sheet Options* groups. We look at most of these tools in the exercises.

Exercise 2: Print Area

Unless you specify otherwise, Excel will print as much of the worksheet as is necessary to show all cell entries and all objects (charts, shapes, etc.). One way to restrict the amount that is printed is to set the Print Area.

(a) Open Sheet6 of Chap2. where we did the van der Waals calculations. For the purpose of this exercise let us assume we wish to print only the first 15 rows. Select A1:H15.

(b) Use the command *Page Layout | Page Setup | Print Area* and click *Set Print Area* (see Figure 3.2).

(c) Now use *Print Preview* to see the effect. You will find that only rows 1 through 15 are in the print preview as requested. Close *Print Preview* and note the dotted page break line just below row 15.

(d) We need to clear the print area setting for the next exercise. Use *Page Layout | Page Setup | Print Area* and select Clear Print Area.

In Excel 2007 it is possible to have multiple print areas for a single worksheet. Try this for yourself: select A1:H10 and set the print area. Select A15:H18 and use *Page Layout | Page Setup | Print Area;* this time there is a third option, Add Print Area. If you click this and use print preview, you will find one page will have A1:H10 and the second page A15:H18.

The Print Dialog

Next we will look at the Print dialog, which is opened with the command *Office | Print | Print* and is shown in Figure 3.3. If you have used any Windows application before, this will be familiar to you. We have the option of changing the printer, assuming the computer has access to more than one printer, how many copies, and if they are to be collated. The *Print Range* area allows you to print all of the worksheet or only some of the pages. We must qualify this a little: if previously you had set a Print Area, then generally only that range of cells gets printed.

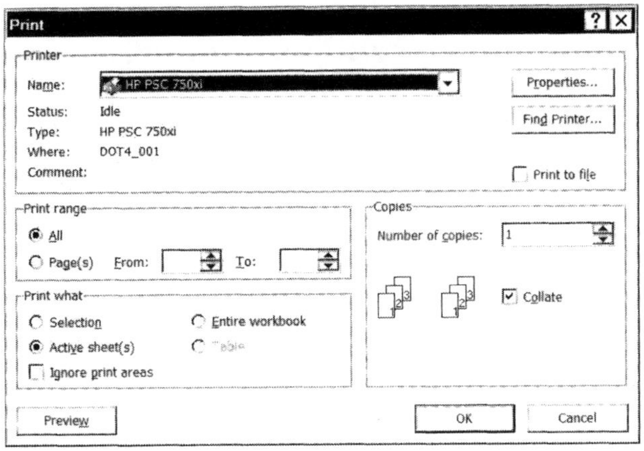

Figure 3.3

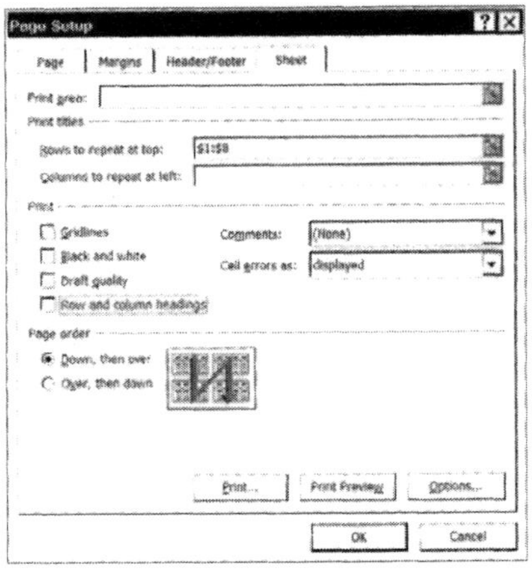

Figure 3.4

The *Print What* area gives us the option of printing just the current selection (a range of cells you had highlighted before opening the Print dialog), the Active (or current) worksheet, or the entire workbook. The *Ignore Print Areas* option explains why the word "generally" was used a few sentences ago; with the option checked, an area that encompasses all the used cells of the worksheet is printed, regardless of any Print Area setting. The *Print Preview* button gives you the option to see the effects of any adjustments you have made before actually committing yourself to using paper. You may wish to experiment with this, but please do not waste paper.

To be economical and "green," the remaining exercises will use *Print Preview* rather than actually printing to paper. You may wish to repeat Exercise 1.1 and add a *Print Preview* command to the Quick Access Toolbar.

Exercise 3: Some Printing Options

So that we can better explore the various print command options, we will extend the van der Waals worksheet as we proceed. First we shall see how to handle the problem of having a worksheet that is just a little too big for one page of print.

(a) Open the Chap2.xlsx file and move to Sheet6, the van der Waals worksheet exercise. To ensure we know how much fits a page, use the *Office | Print | Print Preview* command, which will add page break lines to the worksheet.

(b) Select G8:H18 and drag the fill handle to the right such that there are two columns of data to the right of the vertical page break line. Why did we select G8:H18 and not just H8:H18? Look at the values in row 8; the newly added values increase by 10. This is a useful Excel feature. Had we taken just H8:H18, the temperature values would have been 311, 312...

(c) Select the last two rows and drag down so that there are two or three rows of data below the horizontal page break line. Now we are ready to do some printing experiments.

(d) Use *Office | Print | Print Preview* and you will find the worksheet needs more than one page. Observe what happens when you click the *Next* and *Previous Page* commands. Close *Print Preview*.

(e) The worksheet needs three pages to print all of it, but two of the pages have very little on them. So we may wish to have Excel "squeeze" the material into one page. On the Ribbon, open the Page Layout tab and look at the Scale to Fit group. Change both the *Width* and *Height* setting from *Automatic* to *1 Page* and then do a *Print Preview*. Excel has adjusted the font size to fit the data to one printed page.

(f) Our worksheet fits one page, but it is far off center. On the *Print Preview* dialog, use the *Show Margins* command and then drag the left margin to the right such that the printed material is more or less centered. Does this look better? Close *Print Preview*. Now return the margins to normal using *Page Layout | Page Setup | Margins*. At the end of the exercise we investigate a better way to center the material.

(g) Again looking at the *Page Layout* tab, experiment with the visible Sheet Option commands. You can have the gridlines displayed on the screen or not; likewise, you can have them printed or not. The same is true of the column and row headings. Perhaps if you are displaying a worksheet with a projector, the heading could be a distraction. On the other hand, if you are printing a worksheet for documentation purposes, the heading could be useful.

(h) Suppose our printed van der Waals worksheet was three pages long. We might wish to have the first eight rows displayed on the top of each page. To do this we use *Page Layout | Page Setup | Print Titles*. This opens the dialog shown in Figure 3.4. You may fill in the *Rows to Repeat on Top* box by typing or by using the condense arrow and highlighting the required rows with the mouse. Now use Print Preview to check the effect.

Note that the dialog shown in Figure 3.4 gives you another way of setting all the print options. The dialog may also be opened by using the diagonal arrow ◥ (the dialog launcher) on the *Page Setup* group. If you open the *Margins* tab, and note that this has an option for centering material on the page. The *Page Setup* dialog can also be opened from *Print Preview* but many of its options are disabled.

(i) It is often convenient to have a worksheet printed with headers and footers. Excel 2007 provides a number of ways to do this: (i) you can click the *Page Layout* button on the status bar, (ii) you can open the *Headers and Footers* tab on the Print Setup dialog, or (iii) use the command *Insert | Headers and Footers*. The last method has the most features, including the ability to add the current date and/or time, the file name, page numbers, and so on. The ampersand (&) symbol has special meaning in headers and footers, so if you wish to have one printed you must type in a pair as in *Apples && Oranges*. You are encouraged to experiment with headers and footers.

4

Using Functions

To locate a particular function in Help, start by typing *List of functions*.

Microsoft Excel 2007 provides over 300 worksheet functions, which are divided into 12 groups: Add-in and Automation, Cube, Database, Date and Time, Engineering, Financial, Information, Logical, Lookup and Reference, Math and Trigonometry, Statistical, and Text. To see a full list, open Help and type function list in the search box and then click on *List of worksheet functions (by category)*. To learn more about a specific function, type the name of a function (such as SIN) in the Help search box. In Chapter 8 we shall see that the user may construct user-defined (custom) functions.

If you are familiar with previous versions of Excel, please note that the Analysis Toolpak is no longer required to get the full set of functions.

Functions are always used as part of a formula as in =SIN(A1) or =8+LOG(B1, 2). When a cell contains a formula, the formula bar displays that formula, but the cell generally displays the value produced by the formula. We often use the phrase *the value returned by the formula* for this quantity. In the two examples at the start of this paragraph, the terms SIN and LOG are the *names* of the functions and the quantities A1, B1, and 2 are called *arguments.*

Arguments: Arguments are contained within parentheses, and in the English-language version of Excel they are separated by commas; semicolons are used in other language versions. The number and types of arguments (cell reference, range reference, text, number, Boolean term, etc) depend on the *syntax* (the rules governing its use) of the function.

Depending on the function, the number of arguments may be fixed, variable, or even zero. For example:

zero arguments	=PI( )
one argument	=SQRT(A2)
two arguments	=ROUND(A2, 2)
variable number	=SUM(A1:A10,B3,B4)

The syntax for the square root function is SQRT(number). We may supply the *number* argument with (i) a reference to a cell =SQRT(A2), (ii) a numeric literal =SQRT(45), or (iii) an expression

=SQRT(LOG(A2)*10). In the syntax ROUND(number, num-digits), the *num_digits* arguments can also be any of these, but a simple literal is the most common.

The syntax for SUM is SUM(number1, [number2], [number3], [number4], ...). The square brackets around the last three arguments indicate that these are optional, while the ellipsis (three dots) tells us that we may add more arguments if needed. The *number* arguments can again be a cell reference, a literal or an expression but can also be a range reference as in =SUM(A1:A100). There is nothing in the syntax to tell you this; one needs some basic knowledge of each function to use it correctly. While SQRT(range) would be senseless, SUM(range) is meaningful. At other times, Help is more detailed, and the text of the Help entry gives details on the arguments.

In some circumstances, one may wish to specify an entire row or column as an argument, as in =SUM(A:A), which will sum all of the numbers in column A. Of course, one would not want to put this formula itself in column A; that gives a circular error.

Limit on arguments: When the number of permitted arguments is variable, the maximum number is 255 and the number of characters may not exceed 8192. Note that a range such as A1:A100 counts as one argument, not 100.

Nesting: In the example =ROUND(SQRT(A2)/2, 2), the expression contains another function: we refer to this as nesting. In Excel 2007, nesting may occur to 64 levels.

Error values: If a syntax rule is not followed, the formula will return an error value. If A10 holds a nonnumber, then =SQRT(A1) cannot give a valid result, so it returns an error value (in this case #VALUE!).

If you plan to share a workbook with others using older Excel versions you need to know: (1) the nesting limit was 7, and (2) the number of arguments limit was 30.

The error values are:

#DIV/0!	Division by zero. This would be the result, for example, of =A1/B1 if B1 had a zero value. Note that a blank cell is treated as having a zero value when used in a numeric context like this.
#NAME?	This results when a formula contains an undefined variable or function name.
#N/A	No value is available.
#NULL!	A result has no value.

#NUM!   Numeric overflow; for example, a cell with =SQRT(Z1) when Z1 has a negative value

#REF!   Invalid cell reference. This can be caused by deleting a row or column that is referred to in a formula, or copying a formula inappropriately. For example, trying to drag the formula =A1 in F2 to F1.

#VALUE!   Invalid argument type. For example, a cell with =LN(Z1) when Z1 contains text would return this error.

While not a true error value, we should also mention:

Column is set too narrow for the value/format used in a cell.

When a cell having an error value is referenced in the formula of a second cell, that cell will also have an error value. A worksheet with an error value needs attention. The only exception is #N/A (note: it alone has no exclamation or question mark), which can be taken to mean not applicable or not available. The "error" even has its own function; enter =NA() in a cell and it will display #N/A.

An error you are sure to meet once or twice is the circular reference error. A formula cannot contain a reference to the cell address of its own location. For example, it would be meaningless to place in A10 the formula =SUM(A1:A10). If you try this, Excel displays an error dialog box with *Cannot resolve circular reference*. If you click OK, the Circular Reference tool appears to help you find the source of the problem. An uncorrected circular reference results in a message in the status bar in the form Circular: A10 to warn you of the problem. There are some specialized uses for circular references, but in general these are errors.

Exercise 1: AutoSum Tool

Perhaps the most basic operation done with a spreadsheet is to add a column of numbers. For this reason, Excel has always had an AutoSum tool that can be used to very quickly construct a formula such as =SUM(A1:A6). We shall look at this tool and explore its other features (it can also generate Average, Count, Max, and Min formulas).

Remember the quick way to fill a range with a sequence: type 1 and 2 in the first two cells, select them, and drag the fill handle down to A6.

(a) We will be making the worksheet shown in Figure 4.1. Open a new Excel workbook, and on Sheet1 enter the text shown in A1, C1:C6 and the values shown in A2:A7. Select C1:D1 and use the *Merge and Center* command on the *Home | Alignment* group.

(b) Make A8 the active cell and click the AutoSum tool found in the top left-hand corner of the *Home | Editing* group; look for

The final step in completing a formula is called "committing." We do this using one of: the ⏎ key; the Tab⇥ key; any of the navigation keys ↓, ←, →, ↑ or the check mark ✓ in the formula bar.

the Greek Σ symbol. Alternatively, look in the Function Library group of Formulas on the Ribbon.

(c)   Cell A8 will now contain the formula =SUM(A2:A7) and there will be a mobile dotted line (the *ant track*) around the range A2:A7. Commit the formula.

◢	A	B	C	D
1	Numbers		Results	
2	1		Sum	21
3	2		Average	3.5
4	3		Count	6
5	4		Maximum	6
6	5		Minimum	1
7	6			

Figure 4.1

Note how the AutoSum tool was successful in finding the correct range of addends (numbers to be added). Frequently, the sum value is needed at the bottom of the column, but for this exercise we shall move it.

A little experiment: Replace the 1 in A2 with the word CAT. What happens to the formulas? They just ignore that value and work with the remaining numbers.
But =A2+A3+A4 will give a #VALUE! error.

(d)   Click on A8 and use the shortcut Ctrl+X to cut the formula. Move to D2 and use Ctrl+V to paste the formula.

(e)   Move to A8 and click the disclosure triangle (▼) on the AutoSum command. Select the Average option to get =AVERAGE(A2:A7). Commit the formula and move it to D3.

The active cell need not be directly below (or to the right) of the values to use the AutoSum tool. We will get the Count formula into D4 in one step.

(f)   With D4 as the active cell, open the AutoSum dialog and select Count Numbers. The ant track will be around D2:D3 as these are the closest numbers to the active cell. Use the mouse to select A2:A7 and note how the ant track moves and the formula changes to =COUNT(A2:A7). Now you can commit the formula. The COUNT function returns the number of cells in the range that have a numeric value. The function COUNTA enumerates the cells having nonblank values; it counts text, numbers, dates, and Boolean entries.

(g)　Repeat step (f) to add the Maximum and Minimum formulas.

(h)　Save the workbook as Chap4. xlsx.

The Insert Function Command

fx

Insert
Function

So far we have composed formulas with functions using the AutoSum too. The *Insert Function* tool provides us with access to a greater range of functions. We may access in either with the command *Formulas | Function Library | Insert Function* or with the icon *(fx)* is located to the left of the Formula Bar. Either way opens the dialog shown in Figure 4.2. Within this, you may select the function you wish to use, learn a little of what it does, and insert its arguments. The *Search* box provides some limited help in locating a function based on what you type, but it is not very intelligent. The *Categories* box allows you to filter the list of functions to just one category or to a list of recently used functions. When you become more familiar with the functions you can use one of the other tools in the *Formulas | Function Library* group to locate a function based on its category.

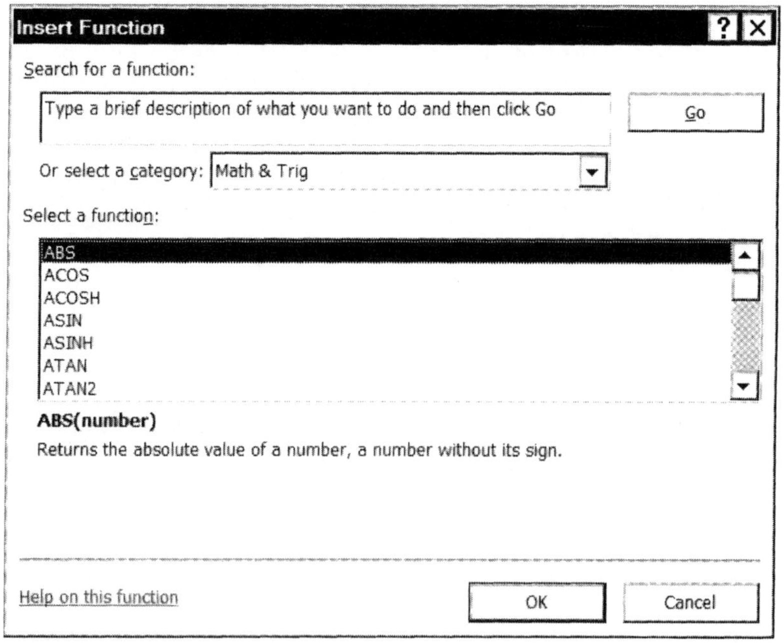

Figure 4.2

Exercise 2: Computing a Weighted Average

For the purpose of this Exercise, let us imagine a student has measured the voltage of a battery many times and recorded his results in a table such as that shown in columns A and B of Figure 4.3. His objective is to compute the weighted average of the results. He needs to calculate

$$\overline{V} = \frac{\sum V_i \times n_i}{\sum n_i}$$

In the course of this Exercise and the next one, we shall compute this value in several ways.

	A	B	C	D	E	F
1	Voltage (V)	Observations (n)	V × N		Sum of Observations	35
2	1.2	1	1.2		Sum of V × N	55.1
3	1.3	3	3.9		Average V	1.574286
4	1.4	5	7.0			
5	1.5	7	10.5		Sumproduct	55.1
6	1.6	8	12.8		Average V	1.574286
7	1.7	5	8.5		Rounded	1.57
8	1.8	3	5.4			
9	1.9	2	3.8			
10	2	1	2.0		Average V	1.57

Figure 4.3

Another experiment: After step (b), delete A6:C10. Type 1.6 in A6 and 8 in B8. When you tab to C6, Excel will automatically add the formula. This is the *Extend Data* feature. Help says it works when at least three of the previous cells have the same formula. The author finds that four repeats seem to be needed.

(a) On Sheet2 of Chap2.xlsx, type in the text shown in A1:C1 and in column E. Use Alt+0215 to make the multiplication sign in C1. Type in the values shown in A2:B10. Use the *Center* command on the *Home | Alignment* group with A1:B10 selected. Use the Border command on the *Home | Fonts* group to add the borders as shown. Gridlines are removed using the tool found in the *Page Layout | Sheet Options* group.

(b) Enter the formula =A2*B2 in cell C2 and fill down to C10 by double clicking the fill handle.

(c) The formulas in F1 and F2 are =SUM(B1:B10) and =SUM(C1:C10), respectively. Compute these using the AutoSum tools as in step (f) of Exercise 1 but using the Sum option. The shortcut to =SUM(is Alt+=.

(d) Compute the average in F3 with =F2/F1.

We will complete the exercise by demonstrating Excel's SUMPRODUCT function. This powerful function makes column C unnecessary. The function computes the sum of the products of elements in two or more ranges. We will compose a formula with this function using the *Insert Function* command.

Function Arguments

SUMPRODUCT

Array1	A2:A10		= {1.2;1.3;1.4;1.5;1.6;1.7;1.8;1.9;2}
Array2	B2:B10		= {1;3;5;7;8;5;3;2;1}
Array3			= array

= 55.1

Returns the sum of the products of corresponding ranges or arrays.

Array2: array1,array2,... are 2 to 255 arrays for which you want to multiply and then add components. All arrays must have the same dimensions.

Formula result = 55.1

Help on this function   OK   Cancel

Figure 4.4

The functions displayed with Most Recently Used are those functions called with the Insert Function dialog. They do not necessarily include functions you may have typed in manually.

The Collapse and Expand tools

(e) With F5 as the active cell, click the *Insert Function* icon to open the dialog shown in Figure 4.2.

(f) Use the *Math & Trig* option in the *Category* box and then select SUMPRODUCT from the lower window. This will open the *Function Arguments* dialog shown in Figure 4.4. Fill this in as shown by clicking first in the *Array1* box and then dragging the mouse over the range to be used. If the dialog gets in the way use the *Collapse* and *Expand* dialog icons to the right of the text box. Note that the Function Arguments dialog gives information on the purpose of the function and displays the final value when all required arguments have been inserted. Click OK and note that the formula in F5 is =SUMPRODUCT(A2:A10, B2:B10).

(g) The average in F6 is computed with =F5/F1.

(h) To round the result to two decimal places in F7, we will use =ROUND(F6,2). Use the *Insert Function* command to compose this. ROUND is in the *Math & Trig.* category.

(i) Save the workbook.

There is a much more direct way in Excel 2007 to get to the Function Argument dialog (Figure 4.4) when you already know the functions's category and do not need to use the Search tool in the *Insert Function* (Figure 4.2) dialog. All you need to do is open the *Formulas* tab on the ribbon and then select one of the categories from the *Function Library* group (see Figure 4.5).

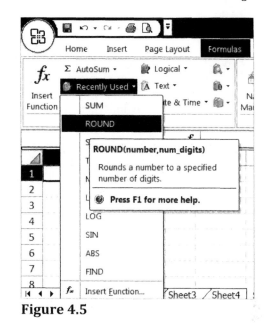

Figure 4.5

In the remainder of the book, it is left to the reader to choose how to insert a function: (i) with *Insert Function*, (ii) with a tool in the *Function Library* group, or (iii) by simply typing.

Exercise 3: Entering Formulas by Typing

We will complete the Worksheet shown in Figure 4.2 by typing a formula in F10. This Exercise (1) shows how to type formulas with functions and (2) demonstrates a two-level nesting formula. The formula we will use will compute the average voltage from the experimental data in columns A and B and round the answer to two decimal places. We shall use a nested formula: =ROUND(SUMPRODUCT(A2:A10,B2:B10) /SUM(B2:B10),2)

(a) In F10 start the formula by typing =R (or =r, since Excel will convert formula names to upper case as it proceeds). A dialog opens, showing a list of all functions beginning with R (see Figure 4.6). As you continue to type, the list is amended until with =ROU there are only four functions showing. You can continue typing or you may click on the ROUND function name to get =ROUND(in the cell.

It would be instructive to make some errors on purpose as you compose this formula to see how Excel reacts.

(b) Continue typing until you have: =ROUND(SUMPRODUCT(.

(c) Use the mouse to fill in the two arguments giving: =ROUND(SUMPRODUCT(A2:A10,B2:B10),2).

(d) Save the workbook.

This is a reasonably complicated formula. It is instructive to use the Evaluate Formula command from *Formulas | Formula Auditing* group to see how the result is obtained. You may also wish to experiment with the Trace Precedents and Remove Arrows commands in the same group. That group also contains the Show Formulas command.

Figure 4.6

Exercise 4: Trigonometry Functions

The following table shows the Excel trigonometric functions together with other functions that are frequently used in conjunction with them. It is essential that the reader remembers that Excel's standard trigonometric functions expect the arguments to be in radians, not degrees, and that the inverse functions return values in radians. While the relationship π *(radians)* $\equiv 180$ *(degrees)* may be used to convert between the two measurements, it is often wise to use the conversion functions: RADIANS and DEGREES.

Functions	Argument	Return value
SIN(n)	any value	-1 to +1
COS(n)	any value	-1 to +1
TAN(n)	any value	With n $=\pi/2\pm\pi$, the value of Tan(n) goes to infinity. Excel gives a large value; =TAN(PI()/2) returns 1.63E+16
ASIN(n)	$-1 \leq n \leq 1$	$-\pi/2$ to $\pi/2$
ACOS(n)	$-1 \leq n \leq 1$	$-\pi$ to π
ATAN(n)	any value	$-\pi/2$ to $\pi/2$

ATAN2(a, b)	1) *a* is the *x-coordinate* of a point, and *b* is the *y*-coordinate 2) ATAN(a/b) and ATAN2(a,b) are equivalent, but *a* can be zero in ATAN2(a,b), giving ±1.5707 (or ± π/2) 3) A positive value indicates an angle counterclockwise from the *x*-axis

◢	A	B	C	D	E
1	Trigonometry Functions				
2	Radians	SIN	ASIN	COS	ACOS
3	0.7854	0.7071	0.7854	0.7071	0.7854
4	1.5708	1.0000	1.5708	0.0000	1.5708
5	2.3562	0.7071	0.7854	-0.7071	2.3562
6	3.1416	0.0000	0.0000	-1.0000	3.1416
7	3.9270	-0.7071	-0.7854	-0.7071	2.3562
8	4.7124	-1.0000	-1.5708	0.0000	1.5708
9	5.4978	-0.7071	-0.7854	0.7071	0.7854
10	6.2832	0.0000	0.0000	1.0000	0.0000

Figure 4.7

(a) Construct the worksheet shown in Figure 4.7. The formula in A3 is =PI()/4. In A4 enter =A3+PI()/4 and copy this down to A10. Do not format the values yet.

(b) The formulas in B3:E3 are =SIN(A3), =ASIN(B3), =COS(A3) and =ACOS(D3), respectively. These are copied down to row 10.

Note the inexact values in some cells; where zero is expected we sometimes get numbers such as 6.13E-17. We will see how to avoid this with the ROUND function.

(c) Go to *Home | Number* and change General in the top box to Number and then use the Increase Decimal tool to have four decimal places showing.

(d) Select A2:E10 and use the shortcut Ctrl+C to copy; move to A12 and use Ctrl+V to paste.

(e) Change A12 to read **Degrees**. Replace the formulas in A13:A20 by the values 45, 90,... 360. Edit the formulas in B13:E13 to read =SIN(RADIANS(A13)), =DEGREES(B13), =COS(RADIANS(A13), and =DEGREES(ACOS(D13)), respectively. Copy B13:E13 down to row 20.

(f) Save the workbook.

Compare the two tables and ensure you understand the use of the functions PI, RADIANS, and DEGREES.

Exercise 5: Exponential Functions

On Sheet3 of Chap4.xlsx, design a worksheet to show that:

(i) =EXP(2) returns e^2.

(ii) =LN(5) returns the natural logarithm of 5.

(iii) =LOG10(5), =LOG(5,10) and =LOG(5) all return the logarithm of 5 to base 10.

(iv) LOG(8,2) returns the value 3, which is the logarithm of 8 to base 2.

Use Help to discover why (iii) is true, that is, the behavior of the LOG function when only one argument is used.

Exercise 6: Rounding Functions

We frequently need to round numbers in calculations. We may be attempting to follow the rules of significant numbers, or we may wish to avoid troubles resulting from binary round-off. Whatever the reason, Excel provides a variety of functions to round or truncate numbers. Remember that formatting is not the same thing. These are shown in the following table.

ABS	Returns the absolute value. =ABS(-12.55) returns 12.55.
CEILING	Rounds a number up (away from zero) to the nearest multiple of significance; cf. FLOOR. =CEILING(1.255, 0.5) returns 1.5.
EVEN	Rounds a number to the nearest even integer. =EVEN(3.25) returns 4.
FLOOR	Rounds a number down (toward zero) to the nearest multiple of significance; cf. CEILING. =FLOOR(1.255,0.5) returns 1.0.
INT	Rounds a number down to the nearest integer; cf. TRUNC. =INT(-5.6) returns -6.
MROUND	Returns a number rounded to the required multiple. =MROUND(6.89,4) returns 8.
ODD	Rounds a number to the nearest odd integer. =ODD(4.25) returns 5.

ROUND	Rounds a number to the required number of places.
	=ROUND(1.378,1) returns 1.4 (one decimal)
	=ROUND(123.56,−1) returns 120 (nearest 10)
	=ROUND(123.56,0) returns 124 (nearest integer)
ROUNDDOWN	Behaves similarly to ROUND but always rounds down.
ROUNDUP	Behaves similarly to ROUND but always rounds up.
TRUNC	Truncates a number to an integer; cf. INT.
	=TRUNC(1.55) returns 1
	=TRUNC(−5.6) returns −5
	INT(x) and TRUNC(x) differ only when the argument is negative.
	TRUNC(x,n) truncates to n decimals places

(a) Open Chap4. xlsx and insert a new worksheet (Sheet5).

(b) Referring to Figure 4.8, enter the values shown in B1, D1, and F1. Name these cells as *x, m,* and *n,* respectively.

(c) Enter the text shown in rows 1, 3, 6, and 8.

(d) In rows 4, 7, and 10, enter the corresponding formulas.

(e) Experiment by changing the numerical values in row 1 to ensure you understand how the functions behave.

	A	B	C	D	E	F	G
1	Value of x	23.456	Value of m	0.5	Value of n	2	
2							
3	ABS(x)	CEILING(x,m)	FLOOR(x,m)	ODD(x)	EVEN(x)		
4	23.456	23.5	23	25	24		
5							
6	INT(x)	TRUNC(x)	TRUNC(x,n)	MROUND(x,m)			
7	23	23	23.45	23.5			
8							
9	ROUND(x)	ROUNDUP(x,n)	ROUNDDOWN(x.n)				
10	23.46	23.46	23.45				

Figure 4.8

Note on Rounding

Most of us round 4.3 to 4 and 4.6 to 5. But what about 4.5? While many would reply 5, others use the round-to-even rule. Thus 4.5 rounds to 4 as does 3.5. Unfortunately, Excel does not provide a function that follows this rule but one can construct a user-defined function (see Chapter 8) that does.

Significant Numbers

There is a very useful formula to round a number to *n* significant digits. You may wish to experiment with =ROUND(A1, A2 -1 - INT(LOG10(ABS(A1)))) where A1 holds the value to be rounded and A2 the number of significant digits required. A literal may replace *A2*. Note that the number may be displayed with extra trailing zeros that are not to be counted as significant. Credit for this formula goes to John Walkenbach.

Some Other Mathematical Functions

As we progress in this book we will meet other mathematical functions such as the matrix functions (MINVERSE, MMULT, MDETERM); functions to generate random numbers (RAND and RANDBETWEEN); various summation functions (SUMXMY2, SUMX2MY2, SUMX2PY2); and so on. The following table lists a few other commonly used functions.

SUMSQ	Returns the sum of the squares of a range of numbers. =SUMSQ(A1:A10)
SUMPRODUCT	Returns the sum of the products of the elements of two ranges—see Exercise 2. Very useful for conditional summations—see Chapter 5.
SQRT	Returns the square root of a positive number.
SQRTPI	Returns the square root of a multiple of π. Thus SQRTPI(2) is equivalent to =SQRT(2*PI())
FACT	=FACT(4) returns the value 4! or $4 \times 3 \times 2 \times 1 = 24$ See also FACTDOUBLE in Help
GCD	Returns the greatest common divisor. =GCD(9, 18, 24) returns 3.
LCM	Returns the largest common multiple. =LCM(9, 18, 24) returns 72.
PRODUCT (limited use)	May be used in place of the multiplication operator; =POWER(A1:A2) and =A1*A2 are equivalent. Can be useful with many numbers are involved as in =PRODUCT(A1:A10).
QUOTIENT (limited use)	May be used in place of the division operator; QUOTIENT(A1,B1) is equivalent to =A1/B1
POWER (limited use)	May be used in place of the exponentiation operator; =POWER(A1,2) is equivalent to =A1^2 .

Array Formulas

All the functions we have looked at so far produce a single result in one cell. There are a number of Excel functions that produce results in a range of cells. When you have finished typing the formula containing one of these functions you must commit (complete) the function with [Ctrl]+[⇧ Shift]+[↵]. None of the other methods of committing a formula will work. A formula that requires this is called *an array formula* in that they return an array of values. We shall be using array formulas throughout the book. In the next chapter we shall see examples of array formulas that produce a single result from an input array.

Exercise 7: The Matrix Functions

Excel includes these functions for working with matrices:

MMULT(A, B)   for matrix multiplication AB
MINVERSE(A)   for finding an inverse A^{-1}
MDETERM   for finding the determinant

We will examine these functions in this exercise and make practical use of them in the next.

For our example of an array formula we shall look at MMULT, which is the function used to multiply two matrices.

	A	B	C	D	E	F	G	H
1	Matrix Functions							
2	Multiplication							
3	Matrix A			Matrix B			Matrix C	
4	2	3		1	2		11	16
5	4	5		3	4		19	28
6								
7	Inverse			Determinant				
8	Matrix A			Determinant of A				
9	-2.5	1.5		-2				
10	2	-1						
11								

Figure 4.9

(a) On Sheet6 of Chap4.xlsx, copy all the entries seen in Figure 4.9 other than G4:H5, A9:B10, and D9.

(b) Select G4:H5 and use the Insert Function tool to create the formula =MMULT(A4:B5,D4:E5) and then holding down [Ctrl]+[⇧ Shift], tap the [↵] key.

The range G4:H5 now holds the matrix C defined by C = AB. If you did not get the expected result, try again. If you are familiar with matrix algebra, you may wish to do the calculation manually.

The MMULT function returns a #VALUE! error if the number of columns of A is not equal to the number of rows of B, or when any cells contain nonnumerical values.

If you happen to forget to use [Ctrl]+[⇧ Shift]+[↵] then you just get a value in the top left corner if you use only [↵] , and #VALUE! errors if you forget the [⇧ Shift].

(c) Click on G4 and note what the formula bar displays: {=MMULT(A4:B5,D4:E5)}. Your formula has been enclosed within braces {} by Excel. This is a "trademark" of array functions.

(d) With G4 still selected, try to delete it with the [Delete] key. You get a message stating that you cannot change just one cell in an array formula.

(e) To find the inverse of matrix A: Select A9:B10, enter =MINVERSE(A4:B5), and again use [Ctrl]+[⇧ Shift]+[↵] to commit the formula.

The MINVERSE function returns a #VALUE! error if the number of columns and rows of A are not equal, or when any cells contain nonnumerical values. Some square matrices cannot be inverted and will return the #NUM! error value with MINVERSE. The determinant for a noninvertable matrix is 0.

(f) The determinant of matrix A is found in D9 with the nonarray formula =MDETERM(A4:B5).

(g) A reader who is familiar with matrix algebra might wish to experiment with =MMULT(A3:B5, A9:B10) in G9:H10

(h) Save the workbook.

Volatility: Calculate Mode

Whenever a value in a cell is changed, Excel normally recomputes every cell that is dependent on the changed value. In a more complex worksheet with many thousands of interrelated formulas, you may see a message such as *Calculating 10%* in the status bar. Usually, the work gets done too quickly for this to be visible.

Some functions get recalculated whenever there is any change made to a worksheet regardless of whether or not the changed cell has an effect on them. Such functions are said to be *volatile*. Some obvious examples are NOW (the current time and date), TODAY (the current date), and RAND (random number). But there are some less obvious ones such as INDIRECT, OFFSET, CELL, and INFO.

If you open a workbook with formulas containing volatile functions and later close it, you will be asked if you wish to save the changes. This message gets displayed even when the user has done nothing to the workbook. The presence of a chart will cause the same message in Excel 2007. Saving is always the best option.

Large workbooks, especially those with many volatile functions, can have long recalculation times. When the workbook gets very complex, the recalculation time can cause a loss of productivity since the user must pause between cell entries. In such cases it is common to set the calculation mode to manual in *Office | Excel Options*. With this setting, the status bar displays *Calculate* whenever a recomputation is needed. Generally, the user presses F9 every so often to have Excel recalculate.

Exercise 8: Solving Systems of Equations

A system of linear equations may be represented in matrix form. Thus the system of two equations:

$$x - 2y \quad = -1$$
$$3x + 4y \quad = 17$$

may be represented by:

$$\begin{bmatrix} 1 & -2 \\ 3 & 4 \end{bmatrix} \begin{bmatrix} x \\ y \end{bmatrix} = \begin{bmatrix} -1 \\ 17 \end{bmatrix} \qquad \text{Equation M}$$

Performing the matrix multiplication in Equation M, we get:

$$\begin{bmatrix} x - 2y \\ 3x + 4y \end{bmatrix} = \begin{bmatrix} -1 \\ 17 \end{bmatrix}$$

When matrix A and matrix B are equal, the corresponding elements are equal. So it follows that $x + 2y = 14$ and $2x - y = 5$; these are the equations with which we started, thereby justifying the statement that we may represent a system of linear equations in matrix form.

Let A represents the matrix of the coefficients, X the matrix of the variables, and C the matrix of the constants.

If the determinant of the matrix is zero, we have a *singular matrix* and no inverse is possible. Hence the system of equations has no solution. The function MDETERM may be used to evaluate the determinant.

Let us write Equation M in the form: $AX = C$

Multiply both sides by A^{-1} giving $A^{-1}AX = A^{-1}C$

But since $A^{-1}A = I$, this becomes $IX = A^{-1}C$

We know that IX = X, therefore $X = A^{-1}C$

From this we see that the value of the X matrix may be obtained by computing A–1C. In Exercise 10 we found the inverse of this matrix, so we may write:

$$\begin{bmatrix} x \\ y \end{bmatrix} = \begin{bmatrix} 0.4 & 0.2 \\ -0.3 & 0.1 \end{bmatrix} \begin{bmatrix} -1 \\ 17 \end{bmatrix}$$

Performing the multiplication gives

$$\begin{bmatrix} x \\ y \end{bmatrix} = \begin{bmatrix} 3 \\ 2 \end{bmatrix}$$

From which we see that x = 3 and y = 2.

This may have left you less than impressed; you could have solved the two simultaneous equations in your head. But the method may be applied to more challenging problems. In this exercise we solve:

$$2x + 3y - 2z = 15$$
$$3x - 2y + 2z = -2$$
$$4x - \ y + 3z = 2$$

The completed worksheet will resemble Figure 4.10.

(a) On Sheet7 of Chap4.xlsx, enter all the all text values . Enter the equation coefficients and constants in A4:C7 and D4:D7, respectively.

(b) Next we compute the inverse (A^{-1}) of the matrix of coefficients. Select the range A10:C12, enter the formula =MINVERSE(A5:C7), and press [Ctrl]+[⇧ Shift]+[↵]. Format the cells to display five places.

(c) The final step to find the solutions is to compute $A^{-1}C$. Select D10:D12, enter the formula =MMULT(A10:C12, D5:D7), and press [Ctrl]+[⇧ Shift]+[↵].

◢	A	B	C	D	E
1		Using Matrix Functions to Solve			
2		a System of Linear Equations			
3					
4		Matrix of Coefficients (A)		Matrix of Constants	
5	2	3	-2	15	
6	3	-2	2	-2	
7	4	-1	3	2	
8					
9		Inverse A^{-1}		A^{-1}C = X	
10	0.19048	0.33333	-0.09524	2	x
11	0.04762	-0.66667	0.47619	3	y
12	-0.23810	-0.66667	0.61905	-1	z
13					
14		Reconstructed equations			
15	46.912	9	2	57.912	
16	70.368	-6	-2	62.368	
17	93.824	-3	-3	87.824	

Figure 4.10

The solutions have now been found. We may wish to check that these agree with the system of equations.

(d) Name the cells D10:D12 as *x*, *y*, and *z*, respectively.

(e) The formulas in row 15 reading from left to right are:
=A5*x   =B5*y   =C5*z   =SUM(A15:C15)
These formulas are copied down to row 17. Save the worksheet.

The values in D15:D17 agree with those in D5:D7, thus confirming that we have solved the system of equations. In Chapter 11 we show the use of this method to solve some practical problems.

Exercise 9: Sum of Diagonal

Sometimes one needs to sum the diagonal elements of a matrix. This is called the *trace* of the matrix. This can be done using the ROW function. In Figure 4.11 we have a small matrix whose diagonal sum is clearly 17. We will look at two ways of finding this with Excel.

The formula in E5 is
=INDEX($A$5:$C$7,ROW(A1),ROW(A1))
while that in F5 is more complex
=INDEX($A$5:$C$7,ROW(A5)-ROW($A$5)+1,ROW(A5)-ROW($A$5)+1).

◢	A	B	C	D	E	F	G
1	Diagonal Sum of Matrix						
2							
3						Safer	
4	Matrix				INDEX	INDEX	SUMPRODUCT
5	2	3	4		2	2	17
6	4	5	7		5	5	
7	6	9	10		10	10	
8					17	17	

Figure 4.11

Why bother with this when the simpler one works? The trouble with the first formula is that if a row is inserted into the worksheet above the matrix, the first formula fails. The second one is "bullet proof."

If just the sum is required, without the separate elements, then use the formula in G5:

=SUMPRODUCT(--(ROW(A5:C7)-ROW(A5)+1= COLUMN(A5:C7)-COLUMN(A5)+1),A5:C7).

The expression --(ROW(A5:C7)-ROW(A5)+1=COLUMN(A5:C7)-COLUMN(A5)+1) evaluates to 1 when the row and column indices are the same, and to 0 otherwise. The double negation converts Boolean FALSE/TRUE to numeric 0/1. You may wish to experiment with the *Formulas | Formula Auditing | Evaluate Formula* tool to see how this formula works.

Financial Functions

As one would expect, Excel offers a very wide range of financial functions. This is not a book on finance, but we shall briefly look at the ones relating to loans and savings.

Financial analysts use a sign convention for the flow of money. Consider the situation were you take out a loan from a bank and make monthly repayments to amortize (pay off) the loan. From your perspective, the initial money coming from the bank to you is considered a positive quantity. The payments you make flow from you to the bank and are considered negative quantities from your viewpoint. The bank, of course, views things the other way around. If you use Excel to evaluate a potential loan, do not be surprised if an Excel calculation differs slightly from the bank's data: there could be banking fees and rounding adjustments.

Lets say you deposit $100 in a savings account and the bank offers an interest rate of 3%. Generally, the advertised rate is a nominal annual rate, but your interest is accumulated monthly. The rate that is used each month is the nominal rate divided by 12. We will assume you leave the original money (the principal) and the interest in the bank for a set period of time.

The $100 deposited today is called the *present value* or *pv*. After a certain number of interest periods (*nper*), the savings may be worth say $125. This is called the *future value* or *fv*. The quantity *payment* or *pmt* is what you pay the bank to amortize the loan or what the bank pays you (deposits into the saving account) as interest earned. The *rate* is the compounding rate. There is just one more quantity: if payments are made at the end of the month then *type* is 0 (this is the default value if you do not enter the *type* argument) and is 1 if payments are made at the start of the month. All these quantities come into play when performing calculations on loans and savings. Excel uses the following equation:

$$pv \times (1 + rate)^{nper} + pmt \times (1 + rate \times type) \times \left(\frac{(1 + rate)^{nper} - 1}{rate} \right) + fv = 0$$

There are Excel functions to compute various quantities. For example the FV function computes the future value. Its syntax is **FV(rate, nper, pmt**, pv, type); remember that bold arguments are required and that the others are optional. We use uppercase for functions and lowercase for arguments.

You plan to deposit $100 a month into a savings plan for five years at a nominal rate of 4%. How much will you have at that time? The answer is found with =FV(4%/12, 5*12, -100). Why is *pv* not used? Because you did not start out with a lump-sum deposit. What would happen if you forgot the negative sign for *pmt*? The result would be negative—you would have taken $100 every month and would be left with a debt!

You win a lottery prize and have to choose between (i) getting $300 a month for six years, or (ii) a lump sum of $6000. Which will you take? We need to look at the present value of each. The present value of a $6000 check is exactly $6000. We can compute the present value for the first option with =PV(5%, 6*12, 300), which gives a result of -$5821. You would need to deposit (hence the negative sign) that amount of money to generate the $300 monthly payment if the bank rate is 5%. Option (ii) is more than this and could generate more than $300 pm if deposited. So option (ii) wins unless there are tax implications.

The function NPER can be used to compute how many periods are needed for a certain scenario, and PMT gives the size of each deposit for another scenario. The function to compute how much is being applied to pay interest is IPMT, while PPMT tells how much is used to pay off the principal. The syntax for RATE (to compute the needed rate for a certain financial scenario) is **RATE(nper, pmt, pv**, fv, type, guess). The last argument may look strange. Look at the financial equation above and you will see it cannot be solved explicitly for *rate*. Excel needs to perform an iterative routine to get an answer. Generally we can omit the *guess* argument. If the successive results of RATE do not converge to within 0.0000001 after 20 iterations, RATE returns the #NUM! error value. Under these circumstances, we can try to assist Excel by giving a guess at the answer.

Problems

1. *When you have tabulated data and no function but need the differential, the following formula may be used.

Forward	Backward	Central
$\dfrac{dy}{dx} = \dfrac{y_1 - y_0}{h}$	$\dfrac{dy}{dx} = \dfrac{y_0 - y_{-1}}{h}$	$\dfrac{dy}{dx} = \dfrac{y_1 - y_{-1}}{2h}$

The voltage drop (V) across an inductance (L) is given by

$$V = L\frac{di}{dt}$$

In Figure 4.12 the central difference formula is used for interior points, and the forward and backward formulas for end-points. What are the Excel formulas in B7:G7?

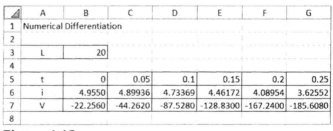

	A	B	C	D	E	F	G
1	Numerical Differentiation						
2							
3	L	20					
4							
5	t	0	0.05	0.1	0.15	0.2	0.25
6	i	4.9550	4.89936	4.73369	4.46172	4.08954	3.62552
7	V	-22.2560	-44.2620	-87.5280	-128.8300	-167.2400	-185.6080
8							

Figure 4.12

2. *To measure the index of refraction μ of a liquid with an Abbé refractometer, a drop is placed between two prisms and a mirror is rotated until the boundary of the light and dark zones align with the cross hairs in a microscope. The index of refraction of the liquid is given by the following equation in

which the angle of rotation is θ and μ_g is the refractive index of glass (1.51). For liquid A, the value of θ was found to be 150° and for B it was 75°. Using a worksheet, find μ for each liquid.

$$\mu = \frac{1}{\sqrt{2}}\left[\left(\mu_g^2 - \sin^2\theta\right)^{1/2} + \sin\theta\right]$$

3. Many engineering applications require *normalizing* an *n*-element vector. If the original vector is V and the normalized one is W, then

$$w_i = \frac{v_i}{\sqrt{\sum_1^n v_1^2}}$$

Giving consideration to the fact that it will be copied across to F4, what is the formula in B4 of Figure 4.13?

	A	B	C	D	E	F
1	Normalized Vector					
2						
3	V	2.3	3.6	5.7	6.8	8.9
4	W	0.173276	0.271214	0.429422	0.512293	0.670501
5						

Problem1　**Problem2**　Probl

Figure 4.13

4. *To fit *n* ordered pairs of data to the equation $y = mx + c$, we can use the formulas

$$m = \frac{n\sum x_i y_i - \sum x_i \sum y_i}{n\sum x_i^2 - \left(\sum x_i\right)^2}$$

$$c = \frac{\sum y_i - m\sum x_i}{n}$$

The Σ in column A of Figure 4.14 is a S which has been formatted with the Symbol font.

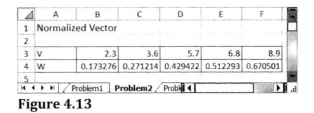

	A	B	C	D	E	F	G
1	Least Squares Fit						
2							
3	x	1	1.5	2	2.5	3	3.5
4	y	5.52	7.67	8.06	9.57	10.55	12.14
5							
6	n	6					
7	Σ(xy)	131.21		slope	2.471429		
8	Σ(x)	13.5		intercept	3.357619		
9	Σ(y)	53.51					
10	Σ(x²)	34.75					

Problem1　Problem2　**Problem3**

Figure 4.14

To fit the data in A3:G4 of Figure 4.14, what formulas would you use in B6:B10 and E7:E8? In Chapter 8 we see much simpler ways to get the same result.

5.   *A trough[1] of length L has a semicircular cross section with radius r. When filled with water to within a distance h of the rim, the volume V is given by

$$V = L\left[0.5\pi r^2 - r^2 \arcsin\left(\frac{h}{r}\right) - h(r^2 - h^2)^{\frac{1}{2}} \right]$$

[1] J. D. Faires and R. Burden. *Numerical Methods,* Brooks/Cole, Pacific Grove, CA, 1998 (page 41).

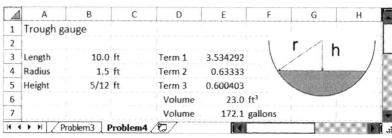

	A	B	C	D	E	F	G	H
1	Trough gauge							
2								
3	Length	10.0 ft		Term 1	3.534292			
4	Radius	1.5 ft		Term 2	0.63333			
5	Height	5/12 ft		Term 3	0.600403			
6				Volume	23.0 ft³			
7				Volume	172.1 gallons			

Problem3   **Problem4**

Figure 4.15

Our task is to make the worksheet shown in Figure 4.15. Typing a long formula is error-prone, so we do it stepwise. What are the formulas in E3:E6? Excel has a CONVERT function, but this does not help here; therefore, Google to find a conversion factor for cubic feet to U.S. or Imperial gallons.

6.   Electrical engineers use a color-coded system to identify the value of a resistor. Thus a resistor with the banding colors green, blue, yellow and gold has a value of 560 kΩ ± 5%. Using only the SUM function, build a worksheet similar to that in Figure 4.16 that will perform the required calculations and check your results on the Internet (for example: http://www.ese.upenn.edu/rca/calcjs.html.) Cell 19 has the custom format [<0.001]##0E+0;[<1000]#0.00;##0E+0 which gives exponents in multiples of three.

We shall expand on this problem below and in later chapters.

7.   Clearly, in the worksheet in Figure 4.16, there should be only one entry in each column header with the *Bar* labels. Use *Data | Data Tools | Data Validation* to give each range a custom validation using the COUNT function such that if the user attempts to enter a second value in a column, the error message in Figure 4.17 pops up. *Hint*: you will need to do this four times, and you will need absolute references .

◢	A	B	C	D	E	F	G	H	I
1	Resistor Four-color Code								
2									
3	R	Color	Bar1	Bar2	Bar3		digit1	digit2	Multiplier
4	0	Black					0	0	0
5	1	Brown					0	0	0
6	2	Red					0	0	0
7	3	Orange					0	0	0
8	4	Yellow			1		0	0	10000
9	5	Green	1				5	0	0
10	6	Blue		1			0	6	0
11	7	Violet					0	0	0
12	8	Grey					0	0	0
13	9	White					0	0	0
14	Tolerance		Bar 4				Digit Digit Multiplier Tolerance		
15	5%	Gold	1						
16	10%	Silver	0						
17	20%	None	0						
18									
19	Resistance		560E+3				Tolerance		5%

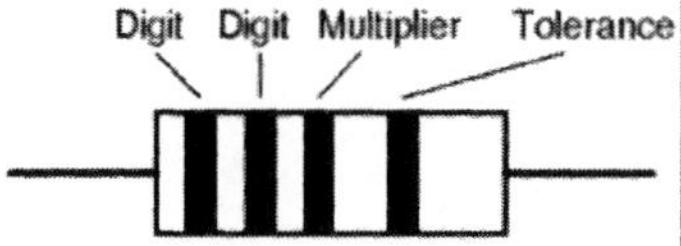

Figure 4.16

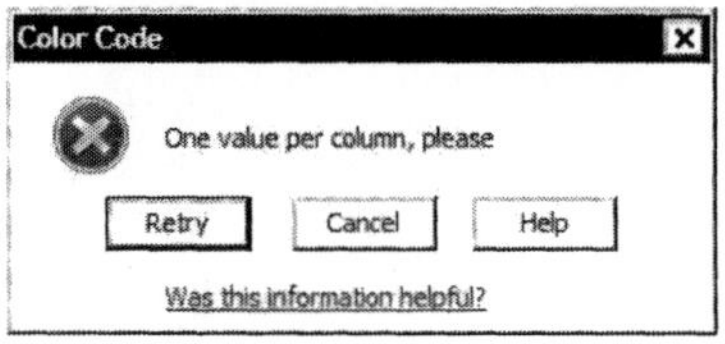

Figure 4.17

8. Create a worksheet to solve the following system of equations using matrix algebra.

$$3x_1 - 4x_2 + 5x_3 + 6x_4 + 2x_5 = 62.5$$
$$x_1 + 2x_2 + 3x_3 + 4x_4 + 5x_5 = 19.5$$
$$6x_1 + 7x_2 - 4x_3 + 2x_4 - x_5 = 15$$
$$5x_1 - 5x_2 + 2x_3 + 5x_4 + 7x_5 = 32$$
$$-3x_1 + 5x_2 + 6x_3 + 2x_4 + x_5 = 16$$

9. You decide to deposit $100 in a savings account on the first day of each month. The bank's nominal rate is 5% per year, but interest is paid monthly. (i) What will be the value of the savings after two years? (ii) Draw up an amortization table as shown in Figure 4.18. Do you get the same answer?

⊿	A	B	C	D
1	Financial			
2				
3	pmt	100		
4	rate	5%		
5	nper	2		
6				
7	month	principal	interest	new principal
8	1	$100.00	$0.42	$100.42
9	2	$200.42	$0.84	$201.25
10	3	$301.25	$1.26	$302.51

Figure 4.18

(iii) The Excel financial functions do not account for the fact that banks will round interest to the nearest penny. Change your amortization table to round the interest. Change the monthly payment to $100,000. How does this effect the difference between the Excel function result and your amortization table? (iv) What answers do you get with the function and the table when deposits are made at month end?

10. Referring to Figure 4.19, PQ is a plane inclined at angle α to the horizontal. A particle is projected from P at a velocity v ft/sec at an angle of β to the plane. Construct a worksheet to compute the range on the inclined plane and the time of flight. What values do you get when α = β = 30° and v = 900 ft/sec using g = 32 ft/s²?

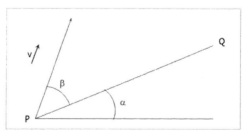

Figure 4.19

Decision Functions

This chapter deals with making decisions or having a cell display a value that is conditional upon what is in one or more other cells. We begin with the logical comparison operators and Boolean functions. We examine the IF function and show how to use SUMIF, SUMIFS, and SUMPRODUCT. Then we examine the table lookup functions LOOKUP, VLOOKUP and HLOOKUP. The chapter concludes with some notes on conditional formatting.

Logical Comparison Operators

The logical comparison operators are used to make tests. We will concentrate on numerical comparisons such as *is the value in A1 greater than 4?*

You may know these operators as relational operators if you have studied computer programming.

The comparison operators are:

=	equal to
>	greater than
>=	greater than or equal to
<	less than
<=	less than or equal to
<>	not equal to

Let A1 hold the value 10, and B1 the formula =A1>10. Since this is untrue, the formula returns the Boolean value FALSE. If we make the formula =A1>=10 then the result will be TRUE.

In computer science, the implicit change of data type is called *coercion*.

In the formula =A1>=10, the A1>=10 part is called a *logical expression*. Logical expressions evaluate to either TRUE or FALSE in Excel. A logical expression has the form:

Expression-1 Logical-operator Expression-2

It can be useful to have logical expressions evaluated to 1 or 0. Excel treats the Boolean values as 1 and 0 when combined with mathematical operations. Following the example above, the formula =(A1>10)*1 will return the value 0 while =(A1>=10)*1 returns the value 1. Using two negation operators is a very efficient method to coerce Boolean values to numeric values; we may use a formula such as =--(A1>=10). When a Boolean value is expected, Excel will accept any nonzero numeric value as TRUE and a zero value as FALSE.

Exercise 1: Boolean Functions

The functions AND and OR may be used to test two or more logical expressions, while the NOT function is used to reverse the truth value of a logical expression.

(a) On Sheet1 of a new workbook, enter the values in A1:B4 of Figure 5.1. Use the information in columns D and F to enter formulas in columns C and E.

(b) Save the workbook as Chap5.xlsx.

◢	A	B	C	D	E	F
1	Boolean Functions					
2	a	4	FALSE	=AND(B2>=5,B3>=5)	TRUE	=OR(B2>=5,B3>=5)
3	b	6	TRUE	=AND(B3>=5,B4>=5)	TRUE	=OR(B3>=5,B4>=5)
4	c	10	FALSE	=NOT(B4=10)	TRUE	=AND((B2+B3),10,B4=10)

Figure 5.1

By default, Excel aligns Boolean values (TRUE and FALSE) centered horizontally in their cells.

Note that if in this worksheet you entered the formula =A2>5, the result would be TRUE. Excel would compare the letter *a* (a text data type) with the literal 5 (also a text data type): the ASCII value for *a* is 97, and that for 5 is 53.

There are some common combinations that are useful to know. In the following table, A and B may be expressions or references to cells containing the values TRUE or FALSE. You may wish to experiment with these nested formulas on Sheet1.

Logic	Formula	TRUE returned if
NAND	=NOT(AND(A,B))	*Not both* true
NOR	=NOT(OR(A,B))	*Neither* is true
XOR	=OR(AND(A, NOT(B)),AND(B, NOT(A)))	*Only one* is true

Exercise 2: Practical Example

Scenario for this Exercise: In a manufacturing plant, 10 items are tested every hour. For each item, two quantities (P and Q) are measured; the P value must meet a certain value, while the Q value must not exceed a certain value. Figure 5.2 shows the worksheet we need to find what percentage of our product is up to specification.

(a) On Sheet2 of Chap5.xlsx, enter the text and numbers shown in A1:B16, C2:C3, and C5:E5 of Figure 5.2.

(b) Name C2:C3 as *pmin* and *qmax*, respectively.

	A	B	C	D	E
1	Quality control				
2		pmin	1.25		
3		qmax	0.5		
4					
5	P	Q	P test	Q test	Two test
6	1.70	0.65	1	0	0
7	1.36	0.50	1	1	1
8	1.44	0.40	1	1	1
9	1.57	0.45	1	1	1
10	1.90	0.82	1	0	0
11	1.52	0.32	1	1	1
12	1.23	0.75	0	0	0
13	1.65	0.50	1	1	1
14	1.29	0.36	1	1	1
15	1.15	0.45	0	1	0
16		Pass	80.0%	70.0%	60.0%

Figure 5.2

(c) The formula in C6 is =- -(A6>=pmin) and this is copied down the column; if you remember the shortcut method of double clicking the fill handle, you will need to delete C16.

(d) In D6 we need the formula =- -(B5<=qmax); and in E6 =- -AND(A6>pmin,B6>=qmax). These are to be copied down to row 15. Can you think of a simpler formula in E6?

(e) Row 16 summarizes the results giving the percentage that passed the tests. We might be tempted to use =SUM(C6:C15)/COUNT(C6:C15) in C16 to get the fraction that passed the P-test. But this is just an average, so why not use =AVERAGE(C6:C15) here and corresponding formulas in D16 and E16?

(f) Save the workbook.

The IF Function

The simple, unnested IF function can be thought of like this: If my *test* is true then return *this-value* else return *that-value*. The function returns one of two possible values depending on the outcome of a logical expression—the test. The syntax is:
=IF(logical-test, true-value, false-value)

The logical test is generally a logical expression we looked at above; for example A1>10. In Help, we are told the logical test must return either TRUE or FALSE. An example of a simple IF is =IF(A1>=10, "OK", "Too small").

However, Excel also allows simple arithmetic expressions in an IF test. Consider the expression A1-10; if this evaluates to a nonzero value, Excel treats it as TRUE, only zero is taken as FALSE. The formula =IF(A1=0,"Zero", "Not zero") could be coded as =IF(A1, "Not Zero", "Zero"). But do not try to be too clever with this approach since others may not follow the logic.

For more complicated formulas, we can nest IF statements. We can replace either or both of *true-value* and *false-value* by another IF statement. Some examples are shown below.

Nesting is permitted to 64 levels in Excel 2007, but getting the logic correct with anything this complex is a major achievement! Generally, it is better to look for a solution using one of the lookup functions.

(i) =IF(A2<0, "Negative","Positive")
Returns the text "Negative" if A2 has a value less than 0; otherwise it returns "Positive."

(ii) =IF(B5<>0, A5/B5, "") or =IF(B5<>0, A5/B5, NA())
Here is a way of preventing the #DIV0! error. These formulas will return a blank or #N/A when the divisor is zero, otherwise the division result is returned.
An alternative formula is =IFERROR(A5/B5, "").

(iii) =IF(ABS(A10−B10)<=EPSILON, "Equal","Unequal")
A cell called EPSILON contains a value such as 1.0E-6. Rather than doing a direct comparison of two values, we test if they differ by more than this value.

(iv) =IF(ABS(A10-B10)<= 0.001,1,0)
Returns 1 if ABS(A10−B10) is less than or equal to 0.001, otherwise it returns 0. We could also use the simpler formula =- -(A10-B10<=0.001).

(v) =IF(SUM(A12:A20)>0, SUM(A12:A20), "Error")
If the sum of the range is greater than 0, that value is returned, otherwise the text "Error" is displayed.

(vi) =IF(A1, TRUE, FALSE) or =IF(A1<>0, TRUE, FALSE)
These will return the value TRUE if A1 contains a nonzero value, a formula giving a nonzero value, or the TRUE value. If A1 is empty or has the value 0, the FALSE is returned. A simpler formula would be =A1<>0, and this would be easier to understand.

The IFERROR function is new to Excel 2007. The first argument is the expression you wish to evaluate, and the second argument is the value to be returned when the first expression results in an error.

Next we look at some examples of nesting.

The logical operators may not be used in array formulas.

(i) =IF(A1>10, IF(A1>50, "Big","Medium"), "Small")
     It is clear that if the condition A1 > 10 is false then the outer IF returns "Small". What happens if the condition is true? The inner IF comes into play. When A1 >50, the inner IF returns 'Big', otherwise it returns "Medium."

(ii) =IF(A1>10, IF(A1>50, "Big","Medium"), IF(A1<0, "Negative", "Small"))

     Here both the true-value and the false-value of the outer IF are themselves IF functions. What results when A1 is 0?

The logical functions AND(), OR(), and NOT() may be used within an IF formula.

(i)   =IF(AND(A2>0, A2<11), A2, NA())
      The value A2 is returned if A2 is greater than 0 and less than 11. Otherwise, the function NA() returns the error value #N/A.

(ii) =IF(OR(A2>0, B2>A2/2), 3 , 6)
      Returns the value of 3 if either A2 > 0 or B2 > A2/2. If neither condition is true, the value 6 is returned.

(iii) =IF(NOT(A2=0), TRUE, FALSE)
      This is the same as IF(A2=0, FALSE, TRUE).

(iv) =IF(NOT(OR(A1=1, A2=1)), 1, 0)
      This is a somewhat contrived example. It returns 1 only when both A1 and A2 have a value that is not 1.

Exercise 3: Resistors Revisited

In Exercise 5 of Chapter 2, we developed a worksheet that computed the effective resistance of four resistors in parallel. We had to invent a workaround to allow us to use the worksheet with fewer than four resistors. With the information in this chapter we can improve our work.

(a) Open Chap2.xlsx, select A1:E6 and click the Copy command on the Clipboard group of the Home tab (or use the [Ctrl]+C shortcut). Open Chap5.xlsx and move to Sheet3. With A1 as the active cell, use [Ctrl]+V to paste the copied material.

(b) Select D6:E6 and then click the eraser icon on the Editing group of the Home tab; select *Erase All* to remove both cell entries and formats.

Now we need to change some of the formulas to give us a worksheet as shown in Figure 5.3.

(c) In B4 replace =1/B3 by =IFERROR(1/B3,""). Copy this across to E4. Place a zero value in E4 to see that the new formula no longer gives #DIV0! but an apparently empty cell when the divisor is zero.

(d) We will find the reciprocal of all sums of the four 1/R values with one formula. At the same time we will round the result to the nearest 10. What we need in B6 is =ROUND(1/SUM(B4:E4),-1).

(e) Select B6:C6 and use the Merge and Center tool. Open the Formatting dialog and give this cell the custom format *0 "ohms."* Test your worksheet.

(f) Save the workbook.

	A	B	C	D	E	F	G	H
1	Resistors in Parallel							
2								
3	R	1240	1800	2000	0			
4	1/R	0.000806	0.00056	0.0005				
5								
6	R_e	540 ohms						
7								
8								

Sheet1 / Sheet2 / **Sheet3** / Sheet4

Figure 5.3

(g) If this were a worksheet in the workplace, the user would have no need to see row 4. Furthermore we would want to ensure that the user could not inadvertently change any formulas in that row. Right click the row header 4 and select Hide. Later we will learn how to protect cell B6.

Exercise 4: Quadratic Equation Solver

In this Exercise we design a worksheet to solve a quadratic equation in the form $ax^2 + bx + c = 0$ using the quadratic formula:

$$x = \frac{-b \pm \sqrt{b^2 - 4ac}}{2a}$$

The quantity $b2 - 4ac$ is called the discriminant; its value determines the number (0, 1, or 2) of real roots of the equation.

When this Exercise is completed, the worksheet will resemble that in Figure 5.4.

◢	A	B	C	D	E
1	Quadratic Equation Solver				
2					
3	a	b	c		disc
4	1	0	-9		36
5					
6	Number of real roots		2		
7	Root 1	3	Root 2	-3	

Figure 5.4

For information on imaginary roots see the workbook IMAGROOTS.XLSX on the companion website.

(a) Open Chap5.xlsx. On Sheet4 enter the values and text shown in A1:C4. Select A3:C4 and horizontally center the entries with the button in the *Home | Alignment* group.

(b) With A3:C4 still selected, use the command *Formulas | Defined Names | Create from Selection* to give the cells A4:C4 the names in the cells above them.

(c) Type disc (short for discriminant) in E3. Enter the formula =b*b - 4*a*c_ in E4. Center E3:E4 and create the name *disc* for the cell E4.

(d) Temporarily ignore the entries in row 6.

(e) Type the text shown in A7 and C7. Enter these formulas
B7:   =(-b + SQRT(disc))/(2*a)
D7:   =(-b - SQRT(disc))/(2*a)

(f) Save the workbook Chap5.xlsx.

You now have an operational worksheet. Test it with quadratic equations whose roots you know. What happens if the value of the discriminant is negative? Try the values 1, 3, and 6 for *a, b* and *c*, respectively. Cells B7 and D7 show the error value #NUM! since it is impossible to evaluate the square root of a negative number without entering the realm of imaginary numbers.

The next steps will improve the behavior of the worksheet when the discriminant is negative and present some additional information.

(g) Enter the text in A6 and center this over A6:B6. In C6 enter the formula =IF(disc<0, 0, IF(disc=0, 1, 2)). This returns 0 when the discriminant is negative, 1 when it is zero, and 2 in all other cases.

(h) Replace the text in A7 with =IF(C6=0,"",IF(C6=1,"Double Root","Root 1")). If there is one root, this returns the text "Double Root," if there are two identical roots, it returns "Root 1." When there are no real roots, it returns an empty test string.

(i) We require the formula in B7 to return a root when the discriminant has a zero or positive value, and an empty text string otherwise. We can achieve this by modifying it to read: =IF(disc>=0,(-b+SQRT(disc))/(2*a), "").

(j) Replace the text in C7 by =IF(C6=2, "Root 2", " ") to return the text "Root 2" only when the discriminant has a positive nonzero value.

(k) Modify D7 to =IF(disc>0,(-b-SQRT(disc))/(2*a), "") to return the value of the second root only when the discriminant has a positive nonzero value. Note that in B7 we tested to see if *disc>=0* while in D7 we tested if *disc>0*; this prevents a double root appearing twice.

(l) Make up some simple quadratics whose roots you know; for example $(x-4)(x+3)=0$ gives $x^2 - x - 12 = 0$ and the roots are clearly 4 and –3. Test that $x^2 - 9 = 0$ reports a "double root." Make any required adjustments.

(m) Save the workbook.

Exercise 5: Protecting a Worksheet

Do not use this type of protection for supersensitive information. Many websites offer password breakers.

Imagine that you have developed a worksheet for use by yourself or others to solve real-world problems. It would be wise to guard against accidental changes being made to cells, especially those with formulas. We will use the quadratic solver as an example. We will arrange things such that the user can visit only the cells needed to define the problem: A4:C4.

There are two steps to the process: (i) specify which cells the user may change by unlocking those cells (by default all cells on a new worksheet are locked) and (ii) switch on worksheet protection.

(a) Open Sheet4 of Chapt5.xlsx. Select A4:C4. Use the command *Home | Cells | Format* to open the menu shown in Figure 5.5. We could just click the Lock Cells item (last but one from the bottom), which acts as a toggle to lock and unlock cells. Alternatively we may use the last item Format Cell... and open the Protection tab to reveal the dialog shown in Figure 5.6. Here we will uncheck the Locked box and click the OK button.

(b) Use the command *Home | Cells | Format | Protect Sheet* to open the dialog shown in Figure 5.7. Our objective is just to prevent accidental changes so we will not use a password. Note that we have the option of allowing the user to visit both locked and unlocked cells. If we did not wish the user to see our formulas, we could deselect *Locked* in this dialog. We shall leave all the other boxes unchecked.

(c) Test the worksheet to see that only the unlocked cells (A4:C4) can be changed. Save the workbook.

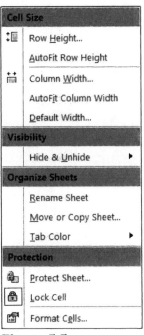

Figure 5.5

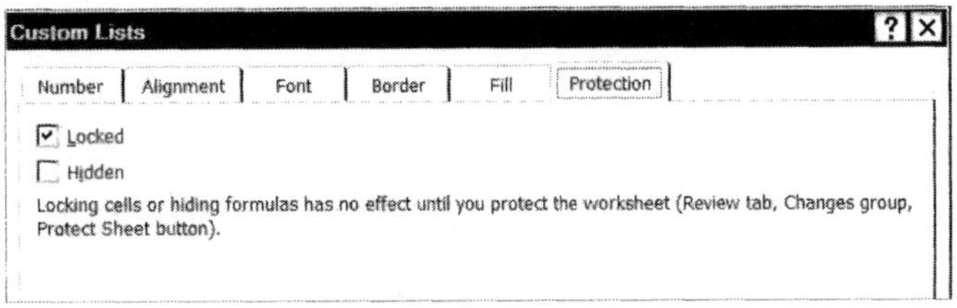

Figure 5.6

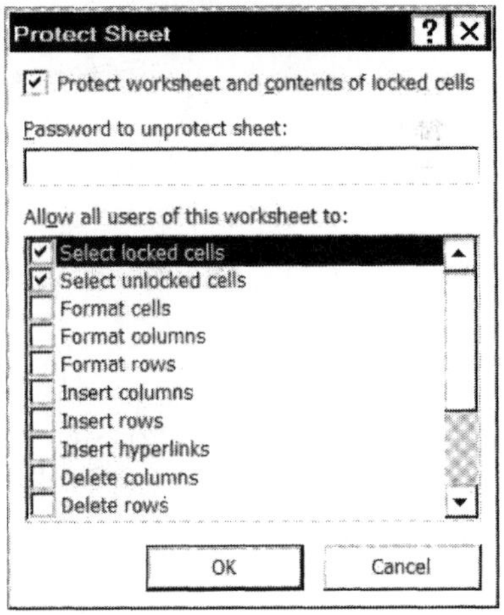

Figure 5.7

Table Lookup Functions

Table lookup functions have a range of uses. Whenever you find yourself composing a multinested IF function, you should consider whether a lookup function would be more appropriate. A vertical table has its headings in a row, while a horizontal one has them in a column. There are no inherent advantages of one over the other.

The functions VLOOKUP and HLOOKUP have similar syntax:

VLOOKUP(***lookup_value, table_array, column_index_num, range_lookup***)

HLOOKUP(***lookup_value, table_array, column_index_num, range_lookup***)

Lookup_value Is the value to be located in the first column of a vertical table (or the first row of a horizontal table). Lookup_value may be either a numeric or text value or a cell reference.

Table_array Is the range reference or name of the table.

Column_index_num (*row_index_num*)
  Is the column (or row) of the table from which the value is to be returned.

Range_lookup Is a logical value (TRUE or FALSE) specifying whether you want an approximate or an exact match. If range_lookup is TRUE or omitted, and there is no exact match, then the function returns the next largest value that is less than the lookup value. If FALSE and no exact match is found, the function will return the error value #N/A. If lookup_value is less than the lowest value in the first column (first row with HLOOKUP), the function returns the #N/A error value.

There is also the LOOKUP function; see Exercise 8.

The MATCH function returns the relative position of a lookup_value in an array. Its syntax is: MATCH(***lookup_value, lookup_array,*** *match_type*). The first two arguments have the same meaning as above. Use 1 for match_type when the table is sorted in ascending order, and you wish to find the largest value that is less than or equal to lookup_value. Use 0 when you needed an exact match; the table need not be sorted. Use –1 when the table is sorted in descending order, and you wish to find the smallest value that is greater than or equal to lookup_value. When *lookup_value* is nonnumeric, MATCH, VLOOKUP, and HLOOKUP are not case sensitive. MATCH may also be used with wildcards.

The INDEX function returns an element from an array and has two forms. The syntax of the first form is INDEX(***array***, *row_num*, *column_num*). Thus =INDEX(A1:C10, 2, 3) returns the value at the intersection of row 2 and column 3 of the table A1:C10. In this example, it returns the value from cell C2.

Exercise 6: A Simple Lookup

For the purpose of this exercise, a geologist wishes to grade some ore samples based on their rare metal content. Ore with 50 to 59 ppm is to be given a low grade: from 60 to 79 merits a medium ranking, from 80 to 99 is considered high, and anything above that is very high. Our completed worksheet will resemble Figure 5.8.

Chapter5.xlsx

	A	B	C	D	E	F	G	H
1	Vertical table lookup example				Lookup Table			
2					ppm	Grade		
3					50	low		
4	Site	ppm metal	Grade		60	medium		
5	A	75	medium		80	high		
6	A	56	low		100	very high		
7	B	86	high					
8	B	60	medium					
9	C	34	#N/A					
10	C	120	very high					
11								

Sheet3 / Sheet4 / **Sheet5** / Sheet7

Figure 5.8

(a) Open Chap5.xlsx and start on Sheet5. For convenience we will enter the lookup on the same worksheet as the ore data. Type the entries shown in E1:F6.

Cell C9, which displays #N/A, has a small green triangle in its upper left corner. When the cell is selected, a warning tip appears, which if opened gives information on the error value. One option is Ignore Error; use this to hide the triangle.

(b) Enter the text and numbers shown in A1:B10 and C4.

(c) The formula in C5 is =VLOOKUP(B5,$E$3:$F$6,2,TRUE). Since we do not want an exact match, we could have used =VLOOKUP(B5,$E$3:$F$6,2). The $ symbols within the references are, of course, needed to keep the reference to the table unchanged as we copy the formula.

(d) Copy the formula down the column.

(e) Save the workbook.

Here are some "experiments" you may wish to try. Between each one, close the file without saving and reopen it.

(f) Modify the formulas in column C with an IF function such that when the B value is less than 50, the cell appears empty.

(g) Get the same effect by moving E3:F6 down one row and adding new data in E3. You will need to modify the formulas.

(h) Cut the table and paste it on a new sheet. Note how the formulas in column C automatically adjust.

(i) Since the table is (i) sorted and (ii) only two columns wide, the LOOKUP function could be used in place of VLOOKUP. Read Help and make the change.

Exercise 7: A Two-Valued Lookup

In this example our tables have more than two columns, so we need some way of indicating in the VLOOKUP formula which one to use. For this Exercise we shall use MATCH.

Scenario: A nutritionist enters a client's height, frame type, and weight, and the worksheet gives the person's optimal weight and a comment on his actual weight. To keep the Exercise to a reasonable size, we limit ourselves to just male clients. Our final product will resemble Figure 5.9.

(a) Begin by entering the table in E1:H16.

(b) Enter the text shown in A1:A12; and the values in B3:B5. Use custom format # ??/12 in B3 so we can use feet and inches. In B5 use custom format 0 "lbs" and apply this to B11 with the Format Painter. We will treat B12 differently.

(c) Create the following names:
frame =F1:H1; height =E2:E16, and optimal = E2:H16.

(d) The formula in B8 is =MATCH(B4,frame,0)+1. Observe how this works: The L in B4 corresponds to the third column in *frame*; we add 1 since we are working with the table *optimal*.

▲	A	B	C	D	E	F	G	H
1	**Optimal male weight**				height (ins)	S	M	L
2					62	130	134	143
3	Height	5 8/12			63	132	137	145
4	Frame	L			64	134	139	148
5	Weight	154 lbs			65	137	141	152
6					66	139	143	154
7	Match functions				67	141	148	159
8	frame	4			68	143	150	161
9					69	145	154	165
10					70	148	156	170
11	Optimal	161 lbs			71	150	159	172
12	Comment	7 lbs	under		72	154	163	174
13					73	156	168	179
14					74	161	170	183
15					75	165	174	190
16					76	168	179	194

Sheet1 / Sheet2 / Sheet3 / Sheet4 / Shee

Figure 5.9

(e) In B11 we have =VLOOKUP(B3*12,optimal,B8). The value in B3 times 12 gives 68; this value is found in the first column of the *optimal*, and the function returns the corresponding value in the fourth column since B8 evaluates to 4.

(f) The formulas in B12 and C12 are, respectively
=IF(B5=B11,"OK",ABS(B11-B5)&" lbs")
=IF(B5=B11,"",IF(B11<B5,"over", "under"))

(g) Save the workbook.

Change the values in B3:B5, observe the results and ensure you understand how the formulas work.

The A7:C8 entries are there for demonstration; we could have combined the B8 and B11 formula as:
=VLOOKUP(B3*12, optimal, MATCH(B4,frame,0)+1)

Whatever can be done with a lookup function can also be done with a combination of INDEX and MATCH. See Problem 2 at the end of the chapter.

Exercise 8: Conditional Summing

The Excel functions SUMIF and COUNTIF may be used to conditional sum or count a range of values. The values are summed (or the entries are counted) subject to specified criteria being satisfied. In the case of SUMIF, the range to be summed may differ from the range to be tested.

Excel 2007 has the new functions SUMIFS and COUNTIFS where multiple criteria may be specified. In addition, Excel 2007 has =AVERAGEIF and AVERAGEIFS.

Unit	Problem	Downtime
A	P2	9
B	P4	14
A	P1	20
B	P4	26
C	P3	14
C	P6	12
C	P6	14

Figure 5.10

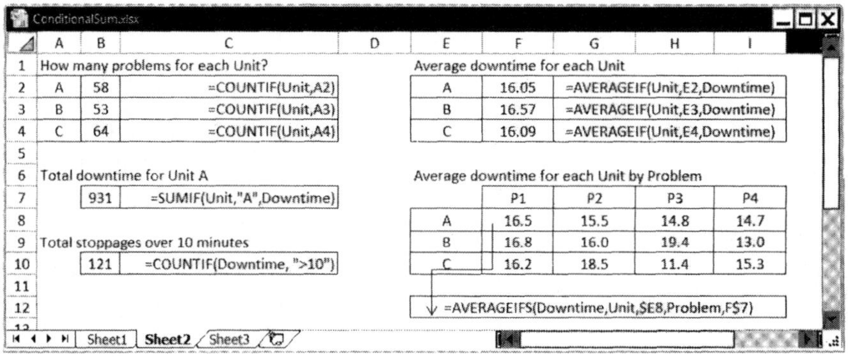
ConditionalSum.xlsx

	A	B	C	D	E	F	G	H	I
1	How many problems for each Unit?				Average downtime for each Unit				
2	A	58	=COUNTIF(Unit,A2)		A	16.05	=AVERAGEIF(Unit,E2,Downtime)		
3	B	53	=COUNTIF(Unit,A3)		B	16.57	=AVERAGEIF(Unit,E3,Downtime)		
4	C	64	=COUNTIF(Unit,A4)		C	16.09	=AVERAGEIF(Unit,E4,Downtime)		
5									
6	Total downtime for Unit A				Average downtime for each Unit by Problem				
7		931	=SUMIF(Unit,"A",Downtime)			P1	P2	P3	P4
8					A	16.5	15.5	14.8	14.7
9	Total stoppages over 10 minutes				B	16.8	16.0	19.4	13.0
10		121	=COUNTIF(Downtime, ">10")		C	16.2	18.5	11.4	15.3
11									
12					=AVERAGEIFS(Downtime,Unit,$E8,Problem,F$7)				

Sheet1 Sheet2 Sheet3

Figure 5.11

Average of Downtime	Column La						
Row Labels	**P1**	**P2**	**P3**	**P4**	**P5**	**P6**	**Grand Total**
A	16.5	15.5	14.8	14.7	14.3	24.6	16.1
B	16.8	16.0	19.4	13.0	18.9	15.3	16.6
C	16.2	18.5	11.4	15.3	17.6	15.7	16.1
Grand Total	**16.5**	**17.0**	**15.4**	**14.4**	**17.0**	**17.5**	**16.2**

Figure 5.12

To get a real appreciation of these functions, one needs a moderately sized data set. Rather than have the user make up data, a file called ConditionalSums.xlsx is available on the companion website. The first few rows of this are shown in Figure 5.10. Each column is named by the label in the top row. Scenario: A manufacturer has three identical productions units (A, B, and C) and has kept a record of the problems they have incurred in a certain time period and how much downtime there was to make each repair/adjustment. Figure 5.11 shows an example of various formulas that could be used to analyze the data.

Pivot tables are very useful for summarizing data; see Exercise 4 in Chapter 6. A sample pivot table made from the data in ConditionalSumx.xlsx is shown in Figure 5.12.

Exercise 9: Array Formulas

In this Exercise we look at constructing array formulas from functions that do not normally need to be treated as array functions. For this we shall first make an array of 20 numbers using the RANDBETWEEN function. We shall use a new method of filling the 20 cells. In the next part of the exercise we will find the sum of the largest five members of the array. In the second part we sum the values regardless of sign. The completed worksheet will have the entries shown in Figure 5.13 (without the text versions of the formulas). The formulas with asterisks are array formulas.

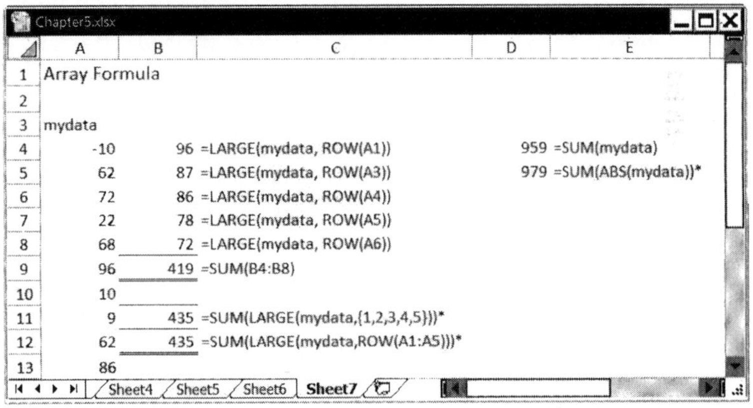

	A	B	C	D	E
1	Array Formula				
2					
3	mydata				
4	-10	96	=LARGE(mydata, ROW(A1))	959	=SUM(mydata)
5	62	87	=LARGE(mydata, ROW(A3))	979	=SUM(ABS(mydata))*
6	72	86	=LARGE(mydata, ROW(A4))		
7	22	78	=LARGE(mydata, ROW(A5))		
8	68	72	=LARGE(mydata, ROW(A6))		
9	96	419	=SUM(B4:B8)		
10	10				
11	9	435	=SUM(LARGE(mydata,{1,2,3,4,5}))*		
12	62	435	=SUM(LARGE(mydata,ROW(A1:A5)))*		
13	86				

Figure 5.13

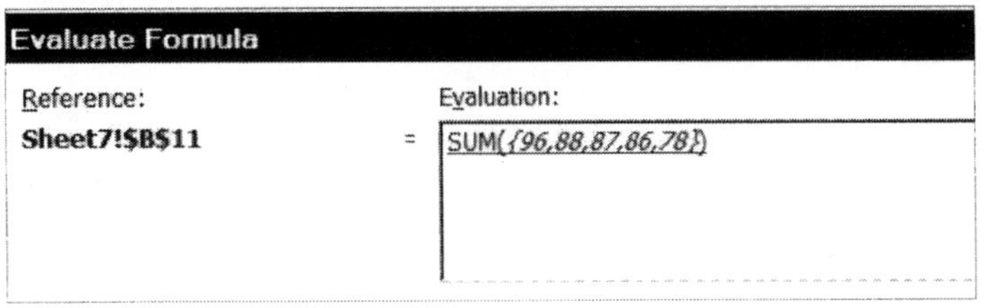

Evaluate Formula

Reference: Evaluation:
Sheet7!$B$11   =   SUM({96,88,87,86,78})

Figure 5.14

(a) On Sheet7 of Chap5.xlsx, enter the text as in rows 1 and 3.

(b) Select A4:A24, type =RANDBETWEEN(1,100) and commit the formula with [Ctrl]+[↵] (note there is no [⇧ Shift]). This is an alternative to typing the formula in one cell and then dragging it down to row 24.

(c) Select A4:A24 (or leave it selected from the previous step). Type mydata into the name box to give our array a name.

We now have 20 random numbers, but these will keep changing since Excel will recalculate the worksheet every time we make a change. Let's turn the formulas into values.

(d) Select A4:A24 and use [Ctrl]+C to copy it.. With the range still selected, right click A4 and open the *Paste Special* dialog. Check the *Values* box and click OK. The formulas are now values. In preparation for the final part of the exercise, change the first value to something negative. Of course, your values will not match those in the figure.

(e) In B4 enter the formula =LARGE(mydata,1). This will return the largest value in *mydata*.

(f) But this is unsatisfactory since it cannot be copied down the column to generate =LARGE(mydata,2). Change the formula to =LARGE(mydata,ROW(A1)). The result is unchanged because ROW(A1) evaluates to 1.

(g) Copy B4 down to B8. Notice how the reference to A1 changes in such a way that we get the top five values in B4:B8.

For those readers with programming experience, an array formula is akin to using a loop structure.

(h) Use the *AutoSum* tool to get the summation of these five in B9.

Finally we are set to demonstrate an array formula. We will now see how we could get the result showing in B9 in one simple formula.

(i) Enter the formula shown for B11 remembering to use [Ctrl]+[⇧ Shift]+[↵] as we learned in Exercise 6 of the last chapter. We get the same result as in B9.
Normally, the function LARGE returns one value as we have seen. But when entered with [Ctrl]+[⇧ Shift]+[↵] it generates an array.

Within B11's formula, the {1,2,3,4,5} part shows how to enter an array of constants.

(j) Select B11 and use the command *Formulas | Formula Auditing | Evaluate Formula*. As you use that dialog, you will see something akin to Figure 5.14.

(k) An alternative formula is used in B12; it is also an array formula.

We have seen an example of an array formula using LARGE; now we look at another with ABS.

(l) Enter the simple formula in D5 and the array formula in D5. Once again we have used a single-values function (ABS) to generate an array; since this is wrapped in a SUM function, we can add these members of the array.

Exercise 10: Conditional Formatting

This chapter has dealt with formulas that return values that depend on a condition (tests, or criteria). In the last exercise we look at conditional formatting: how to give cells a format that depends on a criterion. We may wish values above (or below) a certain value to be displayed in a different color or to be in cells with an eye-catching background fill. Another use is to hide certain values.

We will start by seeing how to hide specified formula values. In Exercise 8 we had a cell that may return #N/A. If we wished this not to appear, we could change the formula in C5 from =VLOOKUP(B5,$E$3:$F$6,2,TRUE) to the lengthy one: =IF (ISERROR(VLOOKUP(B5,$E$3:$F$6,2,TRUE)),"",=VLOOKUP(B5, $E$3:$F$6,2,TRUE)). But conditional formatting might be better.

(a) Select C5:C10 in Sheet5. Use the command *Home | Styles | Conditional Formatting* and in the resulting drop-down menu select *New Rules.* In the next dialog box, click on *Format cells that contain....* and select *Errors* (see Figure 5.15). Click the Format button and on the Font tab set the color to match the cell background color (most likely it will be white). Return to the worksheet and the #N/A value is invisible.

(b) In the same worksheet, perhaps we would like a different background color for the *ppm* depending on the value in the corresponding C cell. Select B5:B10, open the *Conditional Formatting* dialog, and select *Use a formula to determine.....* Complete the resulting dialog as shown in Figure 5.16 and use the Format button to select a background color. Repeat for each possible value in C using different background colors.

Other conditional formatting examples are to be found in the file ConditionalFormat.xlsx on the companion website.

Figure 5.15

Figure 5.16

Exercise 11: SUMPRODUCT

The primary purpose of the SUMPRODUCT function is to compute the sum of the products of the elements of two or more arrays. Thus SUMPRODUCT (A1:A3,B1:B3) evaluated A1*B1 + A2*B2 + A3*B3.

Scenario: A process engineer has taken 25 samples from a product stream and analyzed them for an impurity. The results are tabulated in rows 3 and 4 of Figure 5.17. He needs to compute the average and standard deviation. When computing an average where a measurement x_i occurs n_i times, we speak of a *weighted average*, and it is found with

$$avg = \frac{\sum x_i n_i}{\sum n_i}$$

The numerator is exactly what SUMPRODUCT computes, while the denominator is found with SUM.

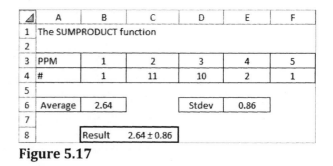

◢	A	B	C	D	E	F
1	The SUMPRODUCT function					
2						
3	PPM	1	2	3	4	5
4	#	1	11	10	2	1
5						
6	Average	2.64		Stdev	0.86	
7						
8		Result	2.64 ± 0.86			

Figure 5.17

(a) Construct a worksheet similar to that in Figure 5.17.

(b) In cell B6 use =SUMPRODUCT(B3:F3,B4:F4)/SUM(B4:F4).

To see how the & operator works, type *apple* in A1 and *pie* in B1. In C1 enter =A1 & " " & B1.

(c) In E6 use =SQRT(SUMPRODUCT((B3:F3-$B$6)^2,B4:F4) / (SUM(B4:F4)-1)) to get the standard deviation. Note how SUMPRODUCT accepts the agreement (B3:F3-$B$6)^2 without requiring that we make it an array function. This is a major strength of the function.

(d) To summarize the results in C8 we use =ROUND(B6,2) & " ± " & ROUND(E6,2). Recall from Chapter 2 that ± is produced with ⎡Alt⎤ + 0177 on the numeric keypad. In this formula the ampersand (&) is used as the concatenation operator—it joins text together.

SUMPRODUCT is also used in ways that the developers may never have considered. These involve counting and summing ranges subject to multiple conditions that COUNTIF and SUMIF cannot manage, being limited to one criterion. The introduction of the Excel 2007 functions SUMIFS and COUNTIFS in Excel 2007 has made some of these "tricks" redundant, but there are still times when SUMPRODUCT out paces these new functions. The primary reason is that SUMPRODUCT allows you to perform operations on the range being summed. Figure 5.18 gives an example.

⊿	A	B	C	D	E	F	G	H
1	The SUMPRODUCT function							
2								
3	Test1	a	a	b	a	a	b	a
4	Test2	x	y	y	x	x	x	y
5	Value	2	3	5	3	5	3	4
6								
7	How many samples have Test1 = a and Test2 = x?					3	3	
8	Sum the values for these two criteria					10	10	
9	Sum the cubes of values for these two criteria						160	

Figure 5.18

We can see that there are three cases with Test 1 = *a* and Test 2 = *x.* The corresponding numbers in row 5 are 2, 3, and 5, which sum to 10. The formulas are as follows.

The double negations in the SUMPRODUCT formulas are used to convert Boolean FALSE/TRUE values to numeric 0/1 values. It is instructive to use the Formula Evaluation tool with the formulas in G7:G9.

	Column F
7	=COUNTIFS(B3:H3,"a",B4:H4,"x")
8	=SUMIFS(B5:H5,B3:H3,"a",B4:H4,"x")
	Column G
7	=SUMPRODUCT(--(B3:H3="a"),--(B4:H4="x"))
8	=SUMPRODUCT(--(B3:H3="a"),--(B4:H4="x"),B5:H5)
9	=SUMPRODUCT(--(B3:H3="a"),--(B4:H4="x"),(B5:H5)^3)

While COUNTIFS and SUMIFS in F7:F8 can replace the SUMPRODUCT in G7:G8, there is no way SUMIFS can sum the cubes of numbers that pass the test.

For a detailed discussion on SUMPRODUCT, see these websites: Bob Phillips at http://www.xldynamic.com/source/xld.SUMPRODUCT.html. J.E McGimpsey at http://mcgimpsey.com/excel/formulae/doubleneg.html.

Problems

[1] B. Carnahan et al., *Applied Numerical Methods*, Wiley, New York, 1969 (page 203).

1. In the *hydraulic jump*,[1] a liquid stream of depth D_1 flowing at velocity v_1, suddenly increases its depth to D_2. Figure 5.19 shows the equation that governs this effect. What formula will you use in E5 that can be copied to H5?

◢	A	B	C	D	E	F	G	H
1	Hydraulic Jump							
2				$D_2 = \dfrac{D_1}{2}\left[\sqrt{1 + \dfrac{8v_1^2}{gD_1}} - 1\right]$ if $v_1 > \sqrt{gD_1}$				
3								
4								
5	D_1	g		v_1	5	10	15	20
6	10	32		D_2	No Jump	No Jump	No Jump	15.8

Figure 5.19

2. *Refer to Figure 5.9 of Exercise 7. We saw that =MATCH(B4,frame,0) tells us which column in the range *frame* matches the frame type entered in B4. Write a formula to find the row position in the range *height* to match the client's height entered in B3. With the existing data in B3:B4, our client's *height* and *type* place him in row 7 and column 3 of the table F2:H16. Write a formula beginning =INDEX that will locate the optimal weight within this. Finally, combine the INDEX formula and the two MATCH formulas into one.

3. *The range A1:A10 in a worksheet contains both positive and negative values, and you wish to sum only the positive ones. Give a formula that will accomplish this.

4. *The range A1:A10 in a worksheet contains both positive and negative values, and you wish find the sum of the squares of only the positive ones. Give a formula that will accomplish this. *Hint*: try either of these:
    (i) SUMPRODUCT, or
    (ii) IF nested inside a SUMSQ as an array formula.

5. *With the same numbers as above, find the average of the squares of the positive values.

6. Construct a worksheet similar to that in Figure 5.20 to make a simple molar mass calculator. Cell C10 uses a SUMPRODUCT formula. Each cell in row 7 uses two IF formulas joined with the concatenation operator &. The first IF gets the symbol, and the next gets the number if it is greater than 1. Then the row 7 cells are themselves concatenated in A10. Hiding rows 3 and 7 would make the worksheet more interesting!

◢	A	B	C	D	E	F	G	H	I	J
1	Molecular Mass Calculator									
2										
3	At. Wt	12.011	1.008	15.999	14.007	30.974	32.065	35.453	79.904	126.904
4	Element	C	H	O	N	P	S	Cl	Br	I
5	Number	2	6				1			
6										
7	Formula	C2	H6				S			
8										
9	Compound		Molar Mass							
10	C2H6S		62.135							

Figure 5.20

7. Refer back to Problem 6 in Chapter 4. This time we will solve the problem without the helper columns. Construct a worksheet similar to Figure 5.21. The cells I11 and I12 each contain formulas that use SUMIF. Alternatively, you may wish to use SUMPRODUCT in your formulas.

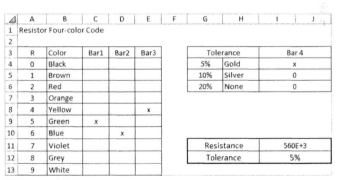

◢	A	B	C	D	E	F	G	H	I	J
1	Resistor Four-color Code									
2										
3	R	Color	Bar1	Bar2	Bar3		Tolerance		Bar 4	
4	0	Black					5%	Gold	x	
5	1	Brown					10%	Silver	0	
6	2	Red					20%	None	0	
7	3	Orange								
8	4	Yellow			x					
9	5	Green	x							
10	6	Blue		x						
11	7	Violet					Resistance		560E+3	
12	8	Grey					Tolerance		5%	
13	9	White								

Figure 5.21

Again we need to protect against having more than one X in a column. Use the same approach as in Chapter 4's problem but with COUNTA rather than COUNT.

8. Rev. Dawn is a recycler; she finds 49 candle stubs and makes exactly seven new candles. These in turn yield seven stubs,

allowing her to make candle number eight from which she later gets a stub. This is illustrated in A3:C7 of Figure 5.22 where we represent the process by $49 \Rightarrow 8R1$.

◢	A	B	C	D	E	F	G	H	I	J	K
1	Recycle										
2											
3	$49 \to 8 R 1$				$79 \to 13 R 1$						
4		stubs	candles	remainder		stubs	candles	remainder		stubs	candles	remainder
5	49	7	0		79	11	2		49	8	1
6	7	1	0		13	1	6		79	13	1
7	1	0	1		7	1	0		72	11	6
8					1	0	1		125	20	5

Figure 5.22

(i)   What formulas are used in F5, G5 and E6?

(ii)   Show that $59 \Rightarrow 9R5$, $67 \Rightarrow 11R1$, $79 \Rightarrow 13R1$, and $88 \Rightarrow 14R4$.

(iii)   From this data, you might conclude that N stubs always yield INT(N/6) candles. You might reason that this is so because, although it takes seven stubs to make a candle, only six get consumed. Show that for N = 72 this is incorrect; under what circumstance does it break down? Algorithms must be fully tested!

(iv)   What formulas are used in J5 and K5?

9.   In the left-hand part of Figure 5.23 we see the solution of a system of equations, while in the right-hand part we see a system with no solution. Refer to Exercise 8 of Chapter 4 and tell what array formula is used in G4:G6.

◢	A	B	C	D	E	F	G		◢	A	B	C	D	E	F	G
1	Matrix								1	Matrix						
2									2							
3		Coefficients		Constants		Soln			3		Coefficients		Constants		Soln	
4	1	2	4	4		x	2		4	1	2	4	4		x	?
5	4	6	9	20		y	5		5	3	6	12	20		y	?
6	2	2	8	-2		z	-2		6	2	2	8	-2		z	?

Figure 5.23

10.   In B11 Exercise 7 we ended up with
=VLOOKUP (B3*12,optimal,MATCH(B4,frame,0)+1)

(i)   Modify it to return "?" if B3 is less than E2 or B4 has an invalid value.

(ii)   Make heights over E16 return "??" and modify B12:C12 to return nothing when B11 is "?" or "??".

6

Data Mining

The purpose of this chapter is to demonstrate how to work with lists containing alphanumeric data. The first topic shows how to import a data list in a non-Excel file. We will then see how to count, sum, and average a numeric column, subject to a criterion relating to a textual column. We will also see the powerful Pivot Table feature. Finally, we will do simple sort and filtering.

The original data were generated by this scenario: Acme Manufacturing has a number of machines in which a particular part must be frequently replaced. There are two sources for this part, Alpha and Beta, each of whom supplies the part in brass, nickel, or stainless steel. The maintenance engineer has kept track of how many hours each part lasts. He keeps his data in a Notepad file; a sample line reads: *Alpha, Brass, 450.* The file Chap6.txt, available from the companion website, contains 100 records.

Exercise 1: Importing a TXT file

We will open the text file and see that Excel provides a tool to parse a record into fields. You may wish to open the file in Notepad to see its contents.

(a) Start Excel and use the command *Office | Open*, make sure the bottom right box of the dialog reads *Text* or *All Files,* and point to the .txt file. Use *Open* or double click its name to bring up the dialog shown in Figure 6.1.

(b) The fields in our text file have a separator between fields (they are delimited) rather than having a fixed width. Having specified *delimited*, click Next to open the second dialog box (Figure 6.2).

(c) Our delimiters are commas, so uncheck the Tab box and check Commas. Note in the preview box the vertical lines that show where the fields begin and end. Click the Finish button.

(d) On a newly inserted row 1, add headers to the columns; Brand, Type, and Hours.

(e) Save your work as an Excel file named Chap6.xlsx.

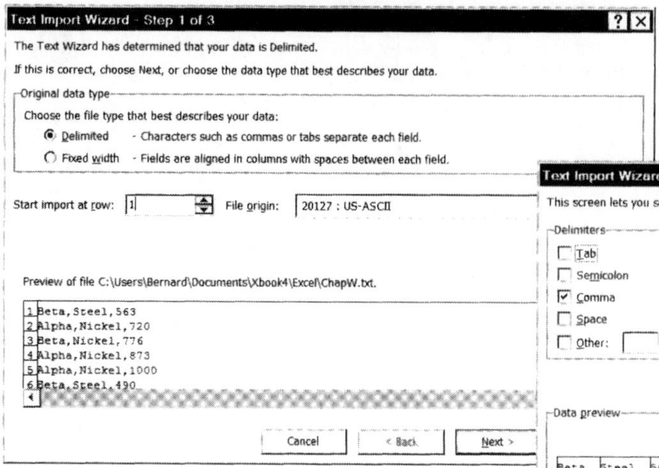

Figure 6.1

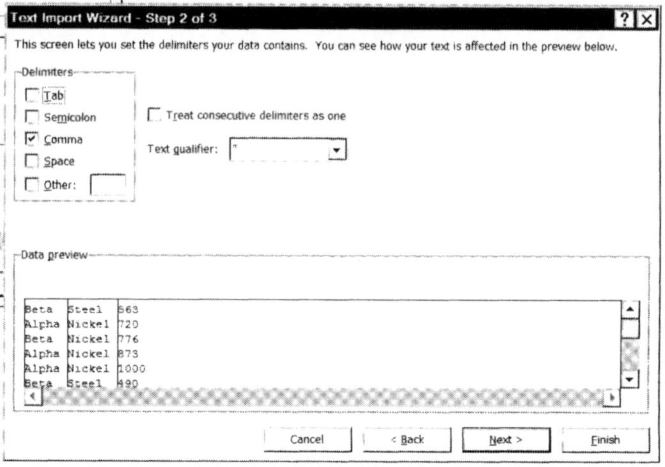

Figure 6.2

Exercise 2: Counting and Summing with Criteria

In this Exercise we look at the functions COUNTIF, SUMIF, AVERAGEIF, COUNTIFS, SUMIFS, and AVERAGEIFS. Note that the last four functions are new to Excel 2007.

(a) On Sheet1 of Chap6.xlsx enter the text shown in Figure 6.3.

(b) Use the following formulas with one criterion:
   F2: =COUNTIF(A:A,E2) to count Alpha items
   G2: =SUMIF(A:A,E2,C:C) to sum Alpha Hours
   H2: =G2/F2 to compute average Hours of Alphas
   I2: =AVERAGEIF(A:A,E2,C:C) as above
   Copy F2:I2 down to next row to find the corresponding Beta values.

Rather confusingly, whereas SUMIF has the *countingrange* as the last argument, SUMIFS has it as the first because it has a variable number of *criteria* arguments.

The new functions allow us to Count and Sum by multiple criteria. We will find, for example, how many brass parts were supplied by Alpha.

(c) Use the following formula to count with two criteria:
   F7: =COUNTIFS($A:$A,$E7,$B:B,F6)
   Use of the absolute and relative references allows us to copy this across to column H and down to row 13.

(d) Use the following formula to sum the hours for each type of part F12: =SUMIFS($C:$C,$A:$A,$E12,$B:B,F11).

Again, this can be copied across and down.

(e)   Average hours may be calculated in F17 with =F12/F7 or in F22 =AVERAGEIFS($C:$C,$A:$A,$E22,$B:B,F21).

(f)   Save the workbook.

⊿	E	F	G	H	I
1		Count	Sum	Average	
2	Alpha	47	28756	611.83	611.83
3	Beta	53	35034	661.02	661.02
4					
5			Count by 2 criteria		
6		Brass	Nickel	Steel	Total
7	Alpha	13	18	16	47
8	Beta	17	12	24	53
9					
10			Sum by 2 criteria		
11		Brass	Nickel	Steel	Total
12	Alpha	7755	11135	9866	28756
13	Beta	12211	7942	14881	35034
14					
15			Average by division		
16		Brass	Nickel	Steel	
17	Alpha	596.54	618.61	616.63	
18	Beta	718.29	661.83	620.04	
19					
20			Average by 2 criteria		
21		Brass	Nickel	Steel	
22	Alpha	596.54	618.61	616.63	
23	Beta	718.29	661.83	620.04	

Figure 6.3

⊿	K	L
1	Frequency	
2	Bin	Count
3	300	1
4	400	14
5	500	16
6	600	11
7	700	18
8	800	18
9	900	10
10	1000	12
11		0

Figure 6.4

Exercise 3: Frequency Distribution

In Exercise 2 of Chapter 16, we make a normal curve from data generated with the FREQUENCY function.

The Hours column has data varying from 300 to 1,000. Perhaps we would like to know how many fall between 300–399, 400–499 etc. We do this with a FREQUENCY function. Many novices have trouble with the Excel Help's use of the term *bin*. A bin is a container; think about sorting red and blue balls into two boxes or two *bins*. We are going to sort our Hours into various bins as shown in Figure 6.4.

(a)   On Sheet1 of Chap6.xlsx enter the text shown in Figure 6.4 and the bin values shown in K3:K10.

(b)   Select L3:L11 and enter =FREQUENCY(C:C,K3:K10) as an array formula. The extra space for the formula is a "safety valve" in case values exceed the last one in the bin. Save the workbook.

Exercise 4: Pivot Tables

The Pivot Table feature is extremely powerful, and we will only be able to scratch the surface of it. Pivot tables are used to summarize data: we will summarize our data as shown in Figure 6.5.

Sum of Hours	Column Labels			
	N	O	P	Q	R
Row Labels	Brass		Nickel	Steel	Grand Total
Alpha		7755	11135	9866	28756
Beta		12211	7942	14881	35034
Grand Total		19966	19077	24747	63790

Figure 6.5

(a) Use the command *Insert | Table | Pivot Table* to open the dialog shown in Figure 6.6. We need to indicate where the data is (A1:C101) and where the output should go (cell N1 of this sheet). Click OK.

Figure 6.7

Figure 6.6

(b) In the Pivot Table Field List dialog (Figure 6.7), we need to check each field that we require in the report. By default, Excel will stack the alphanumeric fields in rows and use the numeric field for the Value. We want Brands to appear as a column, so we must drag the Brands button to Column Labels. You will see the worksheet change as you do this. When ready, close the Pivot Table dialog with the X in its title bar.

(c) We have generated the results shown in Figure 6.6. But what if we wanted average values rather than sums? Click anywhere in the Pivot Table to open the Pivot Table Field List, click on the Sum of Hours button, and choose Value Field Settings to bring up the dialog in Figure 6.8. Now you can not only select Average but you can have a custom name that will appear in cell N1. You may wish to experiment further.

(d) Save the workbook.

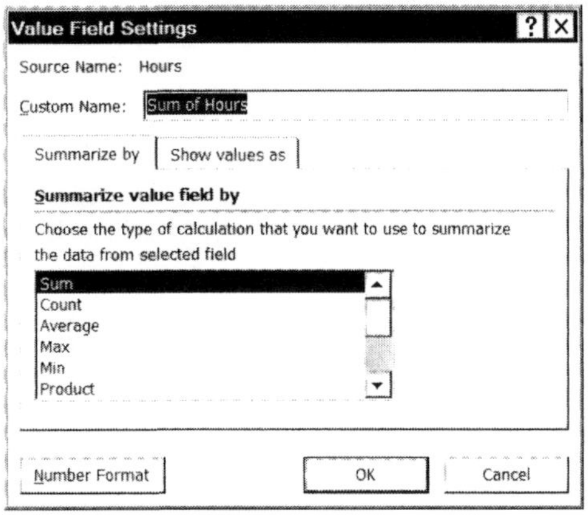

Figure 6.8

Exercise 5: Sorting

Excel 2007 allows you to sort on up to 26 fields. We will sort our data by Brand, Type, and Hours.

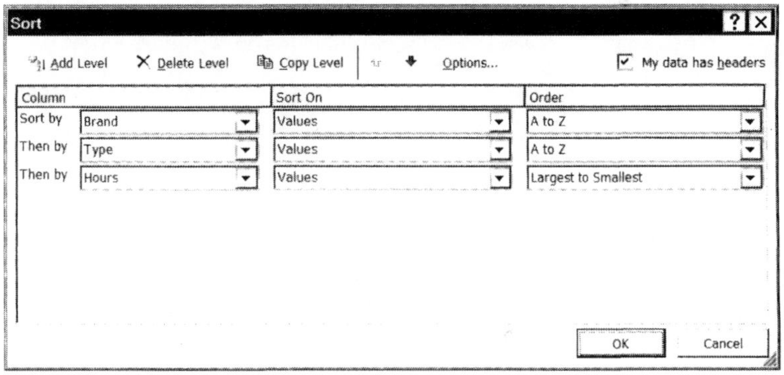

Figure 6.9

(a) To preserve our original data, copy A1:C101 to A1 in Sheet2.

(b) Select any one of the cells in column A. Click on the A → Z icon in *Data/ Sort & Filter* group. The entire table is sorted with column A being alphabetical. Select any one of the cells in column B and repeat the process. You will see that column A is reordered. You will see that you cannot use this method to do a multicriteria sort.

(c) Select a cell anywhere in the table and open the *Data | Sort & Filter | Sort* dialog.

(d) Using the drop-down arrows, sort first on Brand, Values, and A to Z. Then click on *Add Level* and sort on Type, Values, and A to Z and, finally, add another level to sort on Hours, Values, and Largest to Smallest. This is shown in Figure 6.9.

Experiment as much as you like with variations in the sort; using columns in different orders, sorting with increasing or deceasing values, and so on.

Exercise 6: Filtering

Filtering is the process by which we cause only part of our data to be displayed on the worksheet. No data is actually lost; it is just temporarily hidden. Again, this is a powerful feature, so we have space for only a cursory look.

Figure 6.10

(a) Copy the data A1:C101 from Sheet1 to Sheet3.

(b) Select any one of the cells in column A. Click on the Filtering icon in *Data | Sort & Filter* group. This places drop-down arrows in the header row.

(c) Click on the arrow in cell A1 (Brand) to bring up the dialog shown in Figure 6.10. Remove all the check marks except the one for Alpha. Do the same thing in B2, selecting only Brass. The result is as expected, but note that the row headers have been colored to alert you to the fact that some rows are hidden.

(d) Clicking on the Filtering icon in *Data | Sort & Filter* group again will unhide the data.

(e) Save the workbook.

The Sort by Color feature is new to Excel 2007.

Exercise 7: The Excel Table

The Excel Table is new to Excel 2007 although there was a similar feature called List in Excel 2003.

Since any tabular data could be called a "table," this book uses either *Excel Table* or just the word *Table* with a capital *T* to refer to this new feature.

The Excel Table provides another way of filtering data. Tables, however, are much more powerful. We will be able to look at only a few features. Perhaps, because this is a new tool, Help has some particular good items about Tables. One of these is a link to an excellent Microsoft tutorial.

(a) Once again select and copy A1:C101 on Sheet1 and copy it to a new sheet—Sheet4.

(b) With the cursor anywhere in the range A1:C101 on Sheet4, use the command *Insert | Table*. This brings up the dialog shown in Figure 6.11. Excel has correctly located our data, so click the OK button.

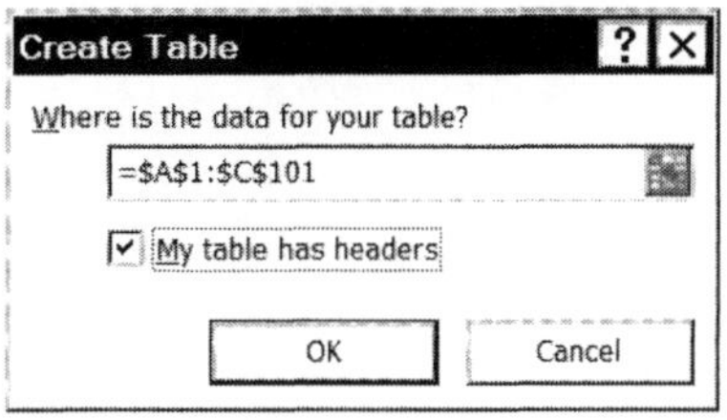

Figure 6.11

As with filtering, the top row now has drop-down arrows. Also some formatting is applied to the Table.

(c) Make C102 (the cell below the last number in the Hours column) the active cell. In *Home | Editing* click the AutoSum tool (Σ). This inserts a formula showing the sum of Hours. But it is not just a sum formula; there is a drop-down arrow that lets you select other functions such as Average and Max.

(d) Now we will add a "calculated column." Type a heading in D1 such as Days. In D2 type =ROUND(C2/24,1) and press ⏎. Excel automatically fills the D column with the formula. If you make this formula with pointing method and clicked on C2, D2's formula would be =ROUND(Table1[[#This Row],[Hours]]/24,1).

(e) Move to A101 and then tab across three times, moving to D101. Tab once more and you are taken to A102: Excel has automatically extended the Table.

This is a major advantage of Tables; they are dynamic. This means that data added to the bottom is automatically incorporated in the Table.

(f) Experiment with filtering. For example, have only the Brass items from Alpha displayed.

(g) Save the workbook.

You may convert the Table back to a simple range by selecting any cell in the Table and clicking the contextual *Table Tool* tab. A new set of items appears on the ribbon. One of these is *Convert to Range*. Experiment with other command Table Tools.

Problems

1. Your company specializes in copper, brass and bronze plate in thicknesses of ¼", ½", ¾", and 1"—so you have 12 products. Every day you receive a file with the orders from your five customers. The first few lines of the file may look like this:

    ```
    Customer,Order Number,Metal,Thickness,Length,Width
    Beta,BVL1000,Copper,0.75,4,6.5
    Zeta,BVL1001,Brass,0.75,11,6.5
    Kappa,BVL1002,Bronze,0.5,4,4.5
    ```

    A file with this type of data is called a *comma delimited* file, and generally it has the extension CSV (from *Comma-Separated Variables*).

(a) Download the file Chap6Data.csv from the companion website. Open it in Notepad to see its contents. Close Notepad and open the file in Excel. Note how Excel imports CSV files automatically putting records (rows) into cells.

(b) Download the file Chap6Pricing.xlsx. This has data on your pricing structure. You charge so much a square foot for each product and an additional cutting fee based on the half-perimeter (length + width).

(c) Use VLOOKUP formulas to calculate the two costs. Wrap this in a ROUND function to get answers to the nearest cent. Finding the metal is easy, but what column to use? Can you see a way to convert the thickness values to appropriate column numbers?

(d) Make the data A1:H101 into an Excel Table. Add Summation formulas to the bottom of the two cost columns. Now you have a worksheet similar to Figure 6.12.

(e) Filter the data to show orders from Alpha. Copy and Paste the data to a new workbook and discover how to e-mail this to yourself from the *Office* tools.

	A	B	C	D	E	F	G	H
1	Custome ▼	Order Numbe ▼	Metal ▼	Thickne ▼	Leng ▼	Wid ▼	$ Me ▼	$ Cutti ▼
2	Alpha	BVL1000	Copper	0.25	13	6.5	756.28	9.75
3	Zeta	BVL1001	Copper	1	11	2.5	895.13	40.50
4	Delta	BVL1001	Brass	0.5	6	6.5	760.50	20.63
98	Kappa	BVL1048	Copper	1	8	3.5	911.40	34.50
99	Delta	BVL1049	Brass	0.5	10	4.5	877.50	23.93
100	Kappa	BVL1049	Bronze	0.75	7	4	994.00	33.00
101	Alpha	BVL1050	Brass	0.75	15	8	3414.00	70.15
102							8608.81	232.46

Figure 6.12

	A	B	C	D	E	F	G	H	I	J	K	L	M	N	O
1	Pivot Table														
2															
3	Count of Customer														
4		0	10	20	30	40	50	60	70	80	90	100	110	120	Grand Total
5	Brass		10	6	2	6	5	3	2	1		2		1	38
6	0.25		2		1	1		1							5
7	0.5		3	1		1	2		1			2			10
8	0.75		4	4		2	1			1				1	13
9	1		1	1	1	2	2	3							10
10	Bronze	2	7	4	4	2	2	4	1			3	1		30
11	0.25		1	2	1		2								6
12	0.5		2	1		1	1					1	1		7
13	0.75		1		2	1		2	1						7
14	1	2	3	1	1		1					2			10
15	Copper		2	2	8	5	3	3	3	2	2		2		32
16	0.25				1	2	2		1	1					7
17	0.5		1		2	1	1	1							6
18	0.75			1	2			1				2			6
19	1		1	1	3	2		2	1	1	2				13
20	Grand Total	2	19	12	14	13	10	10	6	3	2	5	3	1	100

Figure 6.13

(f) Add a column with the header *Area* using =length*width. Note how Excel automatically extends this down the Table. Modify the formula to give the area to the nearest 10 square feet. Do not worry about the odd zero for orders less than $10ft^2$.

(g) Construct a pivot table as shown in Figure 6.13. Use the fields *Metal*, *Thickness* for rows, *Area* for columns, and *Order Number* as the Value Field. Observe the changes when *Order Number* is replaced by *Customer* in this pivot table; explain why the values change.

(h) Create a pivot table showing the average value of the orders from each customer for each metal type.

(i) Download Chap6Generate.xlsx from the companion site to see how the hypothetical data was generated for this problem using the RANDBETWEEN and CHOOSE functions.

2. Prepare a worksheet similar to Figure 6.14: (i) enter the text shown in A1, A3, and F3; (ii) select A3:A503, enter =RANDBETWEEN(1,100), and commit with $\boxed{\text{Ctrl}}$+$\boxed{\leftarrow}$ (no $\boxed{\Uparrow \text{Shift}}$, this time); (iii) copy A3:A500 and use Paste Special to convert formulas to values; and (iv) enter the series 5 to 100 in G5:G23.

◢	A	B	C	D	E	F	G	H
3	MyData					Bin		
4	3		*MyData*			5	*Bin*	*Frequency*
5	6					10	5	25
6	74		Mean	50.56		15	10	24
7	62		Standard Error	1.289261		20	15	21
8	6		Median	49		25	20	25
9	74		Mode	84		30	25	22
10	29		Standard Deviation	28.82875		35	30	28
11	72		Sample Variance	831.0966		40	35	29
12	14		Kurtosis	-1.18836		45	40	23
13	1		Skewness	0.011714		50	45	29
14	41		Range	99		55	50	30
15	95		Minimum	1		60	55	23
16	24		Maximum	100		65	60	23
17	54		Sum	25280		70	65	27
18	87		Count	500		75	70	14
19	43					80	75	26

Figure 6.14

(a) Use a Data Analysis tool to generate the results in columns C and D.

(b) Use a Data Analysis tool to generate the results in columns G and H.

(c) Find which functions give the same results as in column D. Omit those terms with which you are unfamiliar.

(d) Learn how to use the FREQUENCY function to duplicate the data in columns G and H.

7
Charts

Engineers and scientists generally talk about making a *graph,* while other professions make *charts.* Frequently, the graph that the technical person makes is what Excel calls an *XY Scatter chart* where the *x*- and the *y*-values are numeric-ordered pairs. Throughout this chapter we will use the word *chart* since this is the term we must use in Help or when requesting assistance in newsgroups.

We start by making some simple charts and learning how to format various elements in them. We will proceed to make more complex charts, including a combination chart, a chart with error bars, and charts with missing data.

Types of Charts

Figure 7.1 shows the major Excel chart types: column, bar, pie, line, area, and XY (scatter).

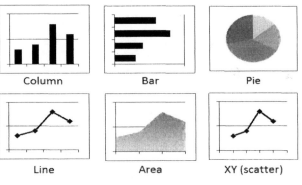

Column   Bar   Pie

Line   Area   XY (scatter)

Figure 7.1

There are also so-called 3D versions of these charts, together with stock, surface, bubble, doughnut and radar. Once you have mastered some basic concepts, you will be able to generate any of these with a little experimentation.

Line and XY Chart

New Excel users often have trouble with the difference between Line and XY charts. The similarity between the two samples in Figure 7.1 is misleading and is somewhat coincidental. To demonstrate this, look at the two charts in Figure 7.2. They were made from the same data but are totally different. The Line chart

If the *x*-values are numbers, you most likely need an XY chart. Look how the Line chart handles the *x*-values in this figure.

treats its *x*-values (the values used to determine the horizontal position) as a *category*. The fact that these were numerical values is totally disregarded This is true of all Line charts except when dates are used for *x*-values. In an XY chart the numerical values of the *x*-data determines the horizontal positions of each data point.

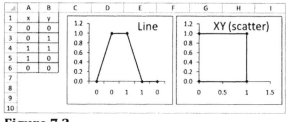

Figure 7.2

The first thing to note is that the use of the name *Line* is misleading. Both Line and XY charts may have lines or no lines joining their data points. A better name would be a *Category* chart. The term *scatter* comes from statisticians; we will stay with *XY*. The second thing to note is that if the *x*-values are to be treated as numbers, then an XY chart is called for.

Comments about Charts

A number of excellent books and articles have been written about charting. Perhaps the most noted are the ones by Edward R. Tufte. His book *The Visual Display of Quantitative Information* (Graphic Press, 1983) is well worth reading. Another is *The Elements of Graphing Data* (Wadsworth, 1985) by William S. Cleveland. Do not be put off by the age of these texts; what they have to say is still relevant today.

The main points raised by these writers may be summarized as:
(i)   Let the chart show the data clearly; do not add extraneous matter—what Tufte calls *chart junk*.
(ii) The chart should not distort the data.
(iii) The chart should show the data at both the broad and the detailed level. The general trend of the data and any fluctuations should be clear.
(iv) The viewer should be drawn to the chart's data, not the method used to construct it.

If one had to summerize the advice in these books, it would be "keep it simple." Of overriding importance is the avoidance of distorting the data. For this reason, they and others deplore the use of the so-called 3D chart—for example, column charts with blocks in place of simple rectangles. Pie charts are similarly criticized for not faithfully depicting their data. So we shall use simple charts in all the Exercises.

Chart Terminology

What follows is a brief introduction to the terminology used by Excel for various chart elements. Many of these elements are optional, as we shall see later. Everything within the borders of the chart is called the *Chart Area*. The *Plot Area* is delineated by the vertical and horizontal axes.

The *Primary x-axis* is generally the lower horizontal border, and the *Primary y-axis* is the left vertical border. The opposite axis is the *Secondary* axis; the qualifier *Primary* is frequently omitted. Of course, in a bar chart the *x*-axis is vertical. The axes are generally divided by *tick marks*. There can be *major* and *minor* tick marks. In Figure 7.3, the horizontal axis has both major and minor tick marks while the vertical has only major ones. *Labels* may be attached to the major tick marks. For example, in Figure 7.3, the scale of the *x*-axis is from 0 to 6. Each primary axis may have major and minor *gridlines*.

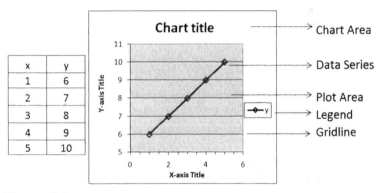

Figure 7.3

A chart may have titles: *chart title; x-axis title,* and *y-axis title*. It may also have a *legend* box. The titles and legend may be dragged to anywhere on the chart, but axes should be close to their own axis line. The size of a title box is determined by its content; a legend box may be resized like a text box.

A chart may have one or more data series. In Figure 7.3 the data is displayed with *markers* (in this case they are diamond shaped), and the points are joined by *lines*. All of these features are optional, but clearly at least one should be used.

Exercise 1: An XY Chart

In this Exercise we will make a chart similar to that in Figure 7.3. We will not have shading (this is called *fill*) in the plot area, nor will we have a legend that is entirely superfluous with only one data series.

(a) Open a new workbook. In A1 of Sheet1 enter a suitable title for the worksheet—something like A Simple XY Chart. Then starting in A3 enter the data shown in Figure 7.3.

(b) We could select all the data in preparation for making the chart, but since there is so little on the worksheet we shall let Excel find the data. Click on any cell in A3:B8. Open the Insert tab on the ribbon. You will see various options in the Chart group. Select Scatter and from the drop-down menu select either of the examples with both lines and markers (the difference between them will be revealed later).

Click on the chart and observe the handles in each corner and in the centers of each side. These may be pulled to change the size of the chart.

You can move the chart by clicking in the outer part of the chart area (avoiding the plot area) and dragging the mouse.

We now have a fairly reasonable XY chart. Double click on the chart. Move the mouse pointer over the chart and note the screen tips that appear. Note also the new item on the ribbon. There is a Chart tab with the groups Design, Layout, and Format. We will look at these later.

The title is the uninspiring letter *y* because Excel by default uses the label of the *y*-values for a chart title. Also we have a redundant legend.

(c) Click on the chart title; a box will appear around it. Type a new title such as Plot of X v Y. That's one way to get a title; let's see another. Again click the chart tile and type = then click on A1 where you have some text and either use the formula bar checkmark or ← to complete the formula =Sheet1!$A$1. Now the title is linked to a cell.

(d) Click on the legend and use the Delete key to remove it.

If Excel does not automatically expand the plot area when the legend is removed, there may be an empty space to the right of the plot area.

(e) Click in the middle of the chart, avoiding the gridlines and the data series. The plot area will be enclosed in a box with small circles in each corner and small squares in the center of each side. Pull the right square to enlarge the plot area.

(f) Note that when the chart is activated (clicked on), colored lines (range finders) appear around the worksheet cells that contain the chart data.

The plot area is outlined by the two axes and the top gridline, but the right side is open. Let's put a line there.

(g) Right click on the plot area; in the resulting shortcut menu select the last item *Format Plot Area*. In the dialog box that appears, open the *Border Color* tab and select *Solid*. There is no OK button on the dialog, just *Close,* so click this. Remember that Ctrl +Z will undo any mistakes you make.

(h) The result may be a line that is too prominent. Repeat the steps above, but open the *Border Style* tab and set the line thickness to 1 point. Save the workbook as Chap7.xlsx.

Our chart now resembles the left-hand chart in Figure 7.4.

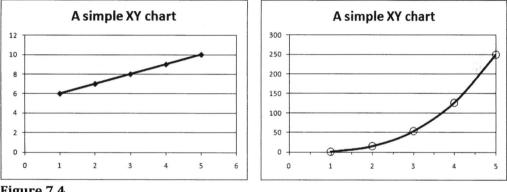

Figure 7.4

Exercise 2: Smooth Lines

In a smooth chart, Excel uses a modified cubic spline to join the markers. Do not confuse this with the line of best fit. We will learn about trendlines in the next chapter.

In step (b) above we saw that there are two XY chart subtypes with lines and markers. In this Exercise we look at that feature.

(a) Open Chap7.xlsx, copy A1:B8 from Sheet1 to Sheet2. Do not copy the chart. Change the data in B3:B8 to: 2, 16, 54, 128, 250. Change the format of the *x*-values so that they display with one decimal place.

(b) Make an XY chart, but this time be sure to select the first subtype that has both lines and markers. Observe how the markers are joined by a smooth line.

(c) Double click the chart; ensure that the *Design* tab of Chart Tools is open and click the *Change Chart Type* command at the far left of the Ribbon. Now each pair of markers is joined by a short line segment. Later we see another way to make this change. Save the workbook.

The appropriateness of smoothing the data or not depends very much on the nature of the data. As a general rule, if the data is expected to follow a mathematical function (example: the pressure of a gas as the volume is changed), then a smooth line may be used. On the other hand, when there is no reason to believe that the two variables are mathematically connected (example: the depth of a lake along a transit line), then a smooth line is generally inappropriate.

Formatting a Chart

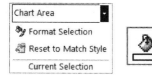

In the Current Selection group, the top command is a drop down box that allows the users to change the selection. This is very useful when two elements (e.g. two data series or a data series and its trendline) are too close to readily permit a correct selection by clicking.

Every element in a chart may be formatted and have its appearance changed. Some examples but not an exclusive list are:

(i)   The color and line style of the border around the plot area can be changed, as can the fill used inside the plot area.
(ii) The color, size, and shape of markers and the color, style, and weight of the line of a data series are all changeable.
(iii) With an axis, one can change the font and the numeric format of labels. The scale can be altered. You can select to have minor and/or major tick marks displayed, and their spacings are alterable.

In fact, if you wish to do anything that seems reasonable then the chances are you can do it. But please recall what was said above about keeping charts simple. A good motto is "just because you *can*, does not mean you *should*." In Excel 2007, Microsoft has made an heroic effort to make it easy to get attractive charts (and to avoid horrid color combinations). However, a lot of the new formatting features are for the commercial user who wants something eye-catching. Technical chart makers are well advised to be less colorful.

There are two ways to start formatting a chart element:
(i)   Click the element; open the *Format* tab in *Chart Tools*; open the format dialog with the *Format Selection* command in the *Current Selection* group at the far left of the ribbon.
(ii) Or, right click the element and use the *Format Element* item at the bottom of the resulting shortcut menu.

Either of these methods opens a formatting dialog for the element. In addition, the Format tab has some formatting commands. There is not sufficient space to discuss every chart formatting feature (indeed, there are books devoted solely to Excel charts), but we will cover enough of the basics to give the user confidence to experiment.

Exercise 3: Formatting the Data Series

If your Excel is set up to use the Office theme (see *Page Layout | Themes*), it is likely that the chart you made in Exercise 2 has blue lines and solid, blue diamond markers. In this exercise we will change the color of both to black, and we will give the markers a hollow circle shape. Then we will format the *x*-axis in various ways.

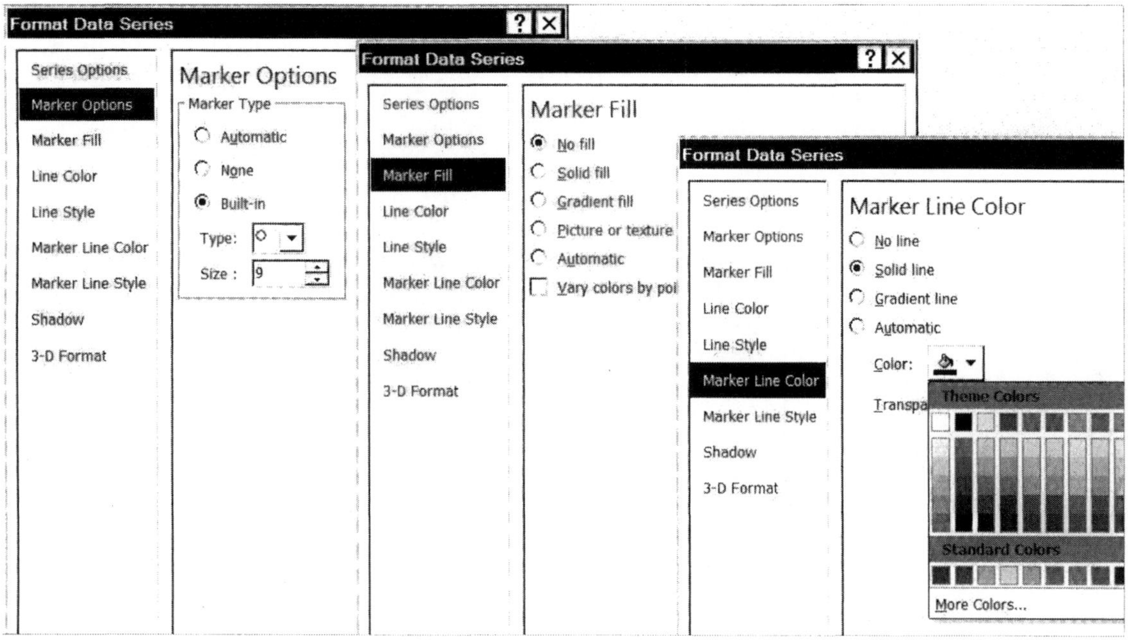

Figure 7.5

The Line Style tab of Format Data Series has an option to make the line smooth or not. You can also change the line thickness from this tab.

The Fill tool (icon of tilted paint can) in the *Home | Font* group can be used to alter the color of markers, columns, etc. You may wish to experiment with this.

(a) Open Sheet2 of Chap7.xlsx. Click on the data series in the chart. Open the Format tab of the Chart Tools and in the Current Section group use *Format Selection* to open the *Format Data Series* dialog box.

(b) In that box open the *Marker Options* tab, select *Built In,* and make the markers circles with a size of 9.

(c) Open the *Marker Fill* tab and select No Fill.

(d) Open the *Marker Line* tab, select *Solid,* and change the color to black.

(e) Open the *Line Color* tab, select Solid, and set the color.

(f) Open the *Line Style* tab and experiment with making a broken or dotted line.

Steps (b) through (d) are illustrated in Figure 7.5. The Line Color dialog is very similar to the Marker Line dialog.

Exercise 4: Formatting an Axis

Use the tools in Home / Font group to alter typeface, font size, etc., of all textual items in a chart.

There are various reasons you might wish to format a chart axis. These include making changes to the scale, the major and minor markers tick marks, the font and/or the number format used for the labels, and the location of the axis relative to the other axis. We look at the first three in this Exercise; the last one we do in Exercise 8.

The chart on Sheet2 of Chap7.xlsx has an *x*-axis with a scale from 0 to 6; it displays only major tick marks, and the labels have an unnecessary decimal place. We will change each of these.

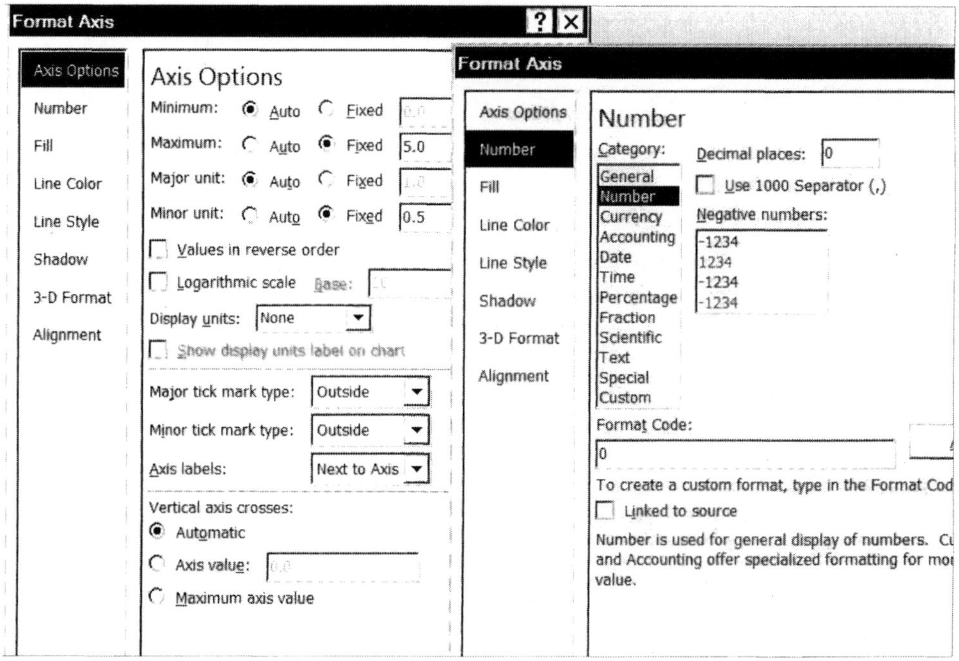

Figure 7.6

By default, the number format for the axes matches that of the data. This, of course, may be overridden.

(a) Open Sheet2 of Chap7.xlsx and click on the *x*-axis. It is best to click on the axis labels. You know when you have hit the mark when a box appears around the *x*-axis.

(b) The ribbon should display the Layout tab of the Chart Tools; select Format Selection. This opens the Axis Option tab of the Format Axis dialog (Figure 7.6). Fix the *Maximum* to 5. Fix the *Minor Unit* to 0.5 and change the *Minor tick mark type* to Outside.

(c) On the same dialog open the Number tab and change the number format to **Number** with 0 decimal places.

(d) Save the workbook.

Exercise 5: Plotting a Function

In this Exercise we plot a function and learn how to extend the chart to include new values. Our finished chart will resemble that in Figure 7.7. We will plot $f(x) = 2x^2-5$ and $g(x) = 3.5x+27$ on the domain –5 to +5.

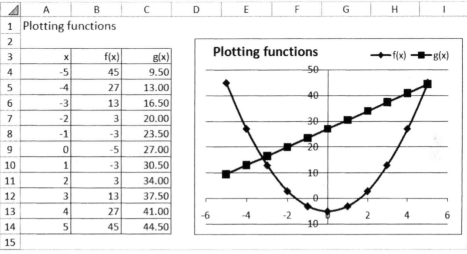

▲	A	B	C	D	E	F	G	H	I
1	Plotting functions								
2									
3		x	f(x)	g(x)					
4		-5	45	9.50					
5		-4	27	13.00					
6		-3	13	16.50					
7		-2	3	20.00					
8		-1	-3	23.50					
9		0	-5	27.00					
10		1	-3	30.50					
11		2	3	34.00					
12		3	13	37.50					
13		4	27	41.00					
14		5	45	44.50					
15									

Figure 7.7

(a) Open Sheet3 and enter all the text, and values shown in A1:A12. Temporarily ignore rows 13 and 14.

(b) Enter the formulas =2*A4^2-5 and =3.5*A4+27 in B4 and C4, respective. Copy them down to row 12.

Here we drag the range finders to extend two ranges in the chart. To use this method to adjust a single data series, begin by clicking that item in the chart then drag the x- and y-rangefinders in two separate operations.

(c) With the mouse pointer anywhere within A3:C12, use *Insert | Charts | Scatter* to make an XY chart with smooth lines.

(d) Add a chart title. Move the legend to the top of the chart area; drag the legend's box to make the two entries side by side. Expand the plot area to fill the chart area.

We have the chart more or less as required. Rows 13 and 14 were ignored, so we may demonstrate a neat way of extending a chart. We want x-values up to 5 with corresponding y-values.

(e) Select A11:C12; pull the fill handle down to row 14. Now we have the needed data, but the chart has not changed.

(f) Click somewhere on the chart, other than one of the data series. Note the range finders (colored lines) around A3:C10. Position the mouse pointer at the bottom between the *x*- and *f(x)*-values. When it changes to a diagonal, two-headed arrow, click and drag the range finders down to row 12. The chart now includes the new data.

(g) Save the workbook.

Exercise 6: More Formatting

Two elements in the chart could be improved. Maybe the chart title's font is too large and the 0 on the *x*-axis is a bit annoying. We will fix both.

An unwanted zero is hidden with the Custom format 0.0;-0.0;" ".

(a) Click the chart title and with the mouse select all the letters in it. Use the font size tool on *Home | Font* to reduce the font size.

(b) Right click on the *x*-axis; use Format Axis to open the formatting dialog. Open the Number tab and select Custom. In the Format Code box type 0.0;-0.0;" ". There are three parts to a custom number format: *code for positive values; code for negative values; code for zero*. We have used a space code for zero to prevent that value from showing.

(c) Save the workbook.

Finding Roots

In Chapter 12 we will show how to use Excel to find roots of f(*x*) =0. Plotting functions can help to find approximate answers to these problems. For example, the roots $2x^2-5 = 0$ are the points where *f*(*x*) cuts the *x*-axis. Of course, this is not very helpful for such a simple function, but the principle holds nevertheless. Similarly, the point of intersection of *f*(*x*) and *g*(*x*) are the solutions to the equations $2x^2-5 = 3.5x+27$, or $2x^2-3.5x-32=0$.

Exercise 7: A Flexible Domain

Frequently, when we plot a function, we are unsure what is the useful range of *x*-values. At other times we may wish to be able to expand a certain part of the chart. This exercise shows one way to do this. We will start by plotting on the domain 0 to 22; but the chart will be flexible enough for us to easily change this to other values.

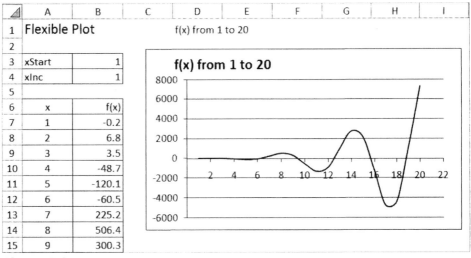

Figure 7.9

(a) On Sheet 4 of Chap7.xlsx, enter the text shown in A1:B6 of Figure 7.9. Select A3:B4 and use *Formulas | Defined Names | Create from Selection* to name B3 and B4.

(b) In A7 enter =xStart, in A8 enter =A7 + xInc. Copy this down to row 26.

(c) In B7 enter =A7^3*SIN(A7)-1/A7. Copy this down to row 26.

(d) Give D1 the formula ="f(x) from " &MIN(A7:A26) &" to " & MAX(A7:A26). The ampersand here is called the *concatenation operator.*

(e) Make the chart as shown. The *x*-axis has the Custom Format 0; -0; " " to hide the nasty zero value of the *x*-axis. The chart title is =Sheet4!D1.

Perhaps we would like to know the approximate roots of f(*x*) in the domain $8 <= x < 14$, but the chart is too dense to read them off.

(f) Change *xStart* to 8 and *xInc* to 0.5. We now see the roots to be approximately 9.5 and 12.5. It is unfortunate that Excel does not use a small value for the minimum *x* value. Save the workbook.

The workbook *FlexibleChart.xlsx* on the companion site uses a slider to allow the smooth changing of *xStart* and *xInc*. The file has instructions on how to add a slider and connect it to a cell value.

Exercise 8: Changing Axis Position

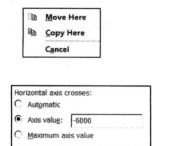

To relocate the *x*-axis, format the *y*-axis, and vice versa!

In the last Exercise, having the *x*-axis along the *y* = 0 line is probably quite appropriate. In other cases one might want the axis at the bottom of the chart. We see how to make a chart resembling Figure 7.10. At the same time we learn how to copy a chart.

The surest way to copy a chart is: Click on chart; use the Copy shortcut Ctrl+C; move to new location; use Paste shortcut Ctrl+V; drag chart to required position. Here are some other ways: (i) Right click in an used part of the chart area to bring up the Move/Copy dialog; click Copy; drag copy into position; (ii) if the right click bring up the shortcut menu use Copy; use the shortcut menu to paste in new position; (iii) hold down Ctrl and click on the chart border and drag to make a copy. Methods (ii) and (iii) suffer from the fact that one seldom clicks in just the right spot to get what is required.

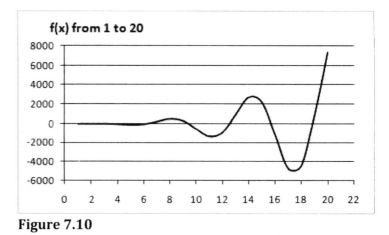

Figure 7.10

(a) Using one of the methods above, make a copy of the chart on Sheet 4.

(b) Format the *y*-axis setting the *Horizontal Axis Crossing* position to –6000. Format the *x*-axis to have a Number format to restore the 0.

(c) Save the workbook.

Exercise 9: XY Chart with Two *Y*-axes

In this Exercise we are presented with three problems: (i) We have two sets of data with different *x*-values; (ii) the *y*-values of the two data sets have very different ranges. One data set has an approximate range of 10 to 26, while the other's range is 0 to 110; (iii) one data set has a missing value. We solve the first problem by making the chart with one data set; then we use Copy followed by

Paste Special to add the second data set. Problem (ii) is solved by using a secondary *y*-axis so that the changes in both *y*-values are clear. For the last problem we do not leave a blank cell but fill it with =NA().

The scenario for this problem is as follows: A researcher has a recording thermometer and a recording light-meter but they are not synchronized. Something interrupted a light-meter reading at time 8.5 hours. The data sets are shown in Figure 7.11.

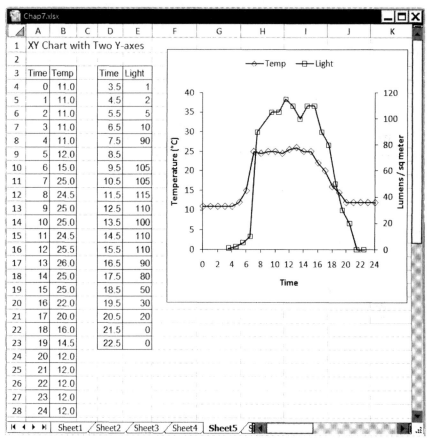

Figure 7.11

(a) Start by entering all the data shown in Figure 7.11 onto Sheet 5. Leave E9 blank for the moment.

(b) Make an XY chart with a nonsmoothed line with the data in A3:B28.

(c) Format the chart as follows; (i) remove the title, legend, and gridlines; (ii) format the *x*-axis to fix the Minimum to 0, the

Maximum to 24, and the Major Unit to 2.

(d) Select D3:E23 and use either Ctrl+C or *Home | Clipboard | Copy* to copy it. Activate the chart by clicking on it. Use *Home | Clipboard | Paste* and select *Paste Special*. Ensure that the *Paste Special* dialog specifies: New Series; *Y*-values in Columns; Series Names in First Row; *X*-values in First Column. Click OK.

We have no data in cell E9, and this is causing a gap in the data series on the chart. There are two ways to fix this.

(e) In E9 enter =NA(), which will display as #N/A . The gap now disappears because the Microsoft Office chart engine ignores this point.

(f) Alternatively, with E9 empty, use *Chart Tools | Design | Data | Select Data* and click the *Hidden and Empty Cells* button to open a dialog where you have three options for how empty cells are to be treated in the chart.

We now have a chart with two data sets. The first thing we need is a secondary *y*-axis. Then we need to separate the two data series vertically by changing the *y*-axes scales. Finally, we shall add some titles.

(g) Open the Format dialog for the second data set (Light) and specify Secondary Axis on the Series Options tab.

(h) Change the primary vertical (the left-hand side *y*-axis) to have a maximum of 40; this will lower the temperature line. Also use a number format with no decimals. Change the secondary vertical axis to have a maximum of 120; this raises the Light line.

(i) Add titles using *Chart | Layout | Labels | Axes Titles*. To get °C use Alt+0176 on the number pad for the degree symbol. From the same ribbon location, restore the legend at the top of the chart. Adjust the plot area to let the titles show clearly.

(j) Save the workbook.

Exercise 10: Control Chart

A control chart generally includes one or more horizontal lines showing a target value; maximum and minimum allowed values; the average value; the average ± the standard deviation (for example, a Levey-Jennings chart), and so on. The technique used here is applicable to all these, provided one is using an XY chart. With Line and Column charts different techniques are needed; see the file *ControlChart.xlsx* on the companion website.

Scenario: The temperature of a chemical process vessel has been measured every hour. A chart is needed showing the hourly values ± standard deviation.

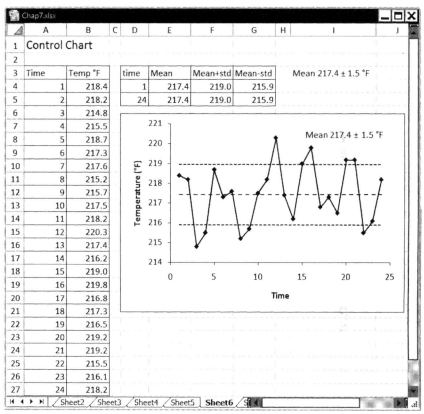

Figure 7.12

Only XY charts allow you to use two data points in this simple way for the control lines. Chart types that use category *x*-values need other techniques; see Problem 4.

(a) Enter the data shown in columns A and B in Sheet6 of Chap7.xlsx (Figure 7.12). Make a nonsmoothed XY chart of this data.

(b) Enter the text in D3:G3. Enter these formulas:

D4   =MIN(A4:A27)
D5   =MAX(A4:A27)
E4:   =AVERAGE(B4:B27)
F4   =E4+STDEV(B4:B27)
G4:   =E4-STDEV(B4:B27)
E5:   =E4 and similar formulas in F5 and G5

There is another way to add a new series to a chart: Open the *Select Data* dialog; click the *Add* button; and fill in the *Edit Series* dialog .

(c) Select D3:G5 and Copy. Activate the chart; use *Home | Paste | Paste Special* as we did in the last exercise.

(d) Format the chart to suit your requirements.

(e) In I3 enter
="Mean "& TEXT(E4,"0.0")& " ± " &TEXT(STDEV(B4:B27),"0.0 °F")

(f) With the chart activated, use *Insert | Text | Textbox*. In the Formula bar type = and click on I3 to generate the formula =Sheet6!$I$3. Resize and position the text box as needed.

Exercise 11: Too Much Data

Imagine you have an instrument whose recordings are captured into an Excel file. You have a very large number of data pairs—too many to make an acceptable chart. We will see how to overcome this problem.

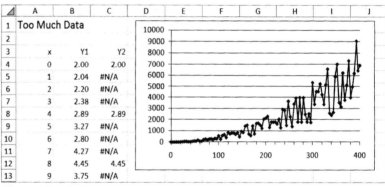

	A	B	C	D	E	F	G	H	I	J
1	Too Much Data									
2										
3	x	Y1	Y2							
4	0	2.00	2.00							
5	1	2.04	#N/A							
6	2	2.20	#N/A							
7	3	2.38	#N/A							
8	4	2.89	2.89							
9	5	3.27	#N/A							
10	6	2.80	#N/A							
11	7	4.27	#N/A							
12	8	4.45	4.45							
13	9	3.75	#N/A							

Figure 7.13

(a) On Sheet7 of Chap7.xlsx enter the text shown in Figure 7.11. In A4 enter 0; click on A4's fill handle and hold down [Ctrl] while dragging down to A404 to fill the range with 1 to 400.

(b) We will make up some *y* data. In B4 enter the formula =(A4/5)^2*(0.5+RAND())+2 and double click B4's fill handle to fill the *y*-range. With B4:B404 still selected, use *Copy* followed by *Paste Special with Values* specified to convert the formulas to values.

(c) We cannot make a clear chart from 400 data points but we can from 40. In C4 enter =IF(MOD(ROW(A4),4)=0,B4,NA()) and copy down the column. This picks out every fourth value from column B. Remember the chart engine will ignore N/A.

(d) Make a chart using A4:A404 and B4:B404. Method 1: Select A4:A404; then, while holding down ⌈Ctrl⌉, select C4:C404; use the command to insert a nonsmooth XY chart. Method 2: with any cell in A4:C404 as the active cell issue the chart-making command; activate the chart and use *Chart Tools | Design | Data | Select Data* to open a dialog where you can delete the first *y* data series.

(e) Save the workbook.

Exercise 12: Large Numbers and Log Scale

The *y*-axis labels in Figure 7.13 give the chart a poor appearance. Figure 7.14 shows a solution.

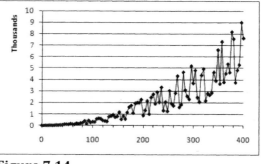

Figure 7.14

(a) Make a copy of the chart on Sheet7 (see Exercise 8). Open the dialog to format the *y*-axis of the new chart; on the Axis Option tab locate the controls to achieve the result required.

(b) As an alternative, have Excel not display the *thousands* text, but use a *y*-axis title with text such as Measurement (in thousands).

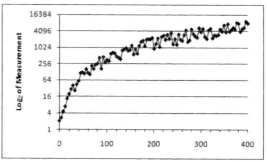

Figure 7.15

Sometimes when you have a large range, a logarithmic scale might help. Excel 2007 allows you to not only make a scale logarithmic, but also to specify the base of the logs. We will use base 2 for no other reason than we can!

It is also possible to format any numeric axis to use a logarithmic scale.

(c) Make another copy of the chart and find, on the formatting dialog, the control needed to make the primary *y*-axis logarithmic to base 2. See Figure 7.15.

Exercise 13: Error Bars

Excel provides many options for adding error bars, but we have space to examine only one. Scenario: You have applied a voltage to a piece of equipment and measured the temperature ten times during an hour before increasing the voltage. Your data is as shown in Figure 7.16. The Temp values are the hourly averages, each Plus value is the recorded maximum positive fluctuation from the average, and Minus is the negative fluctuation.

(a) On Sheet8, enter the data shown in A1:D8 in Figure 7.16. Select A3:B8, make an XY chart, and format it as required.

(b) The steps to add the error bars are: Activate the Chart; use *Toolbars |Layout |Analysis |Error Bars;* click on *More Error Bars Options;* on the Vertical Error Bars tab locate the *Error Amount* box and check Custom; use Specified Value and in dialog box Custom Error Bars, use the mouse to select *C4:C8* for the *Positive Error Values* box and D4:D8 for the Negative Error Values box; click OK and then click *Close* on the *Format Error Bars* dialog.

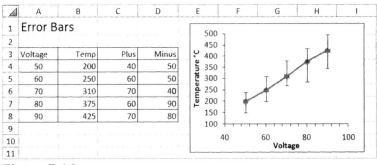

Figure 7.16

(c) As of the time of writing when SP1 was the current upgrade for Excel 2007, this procedure works but incorrectly adds horizontal error bars also. If you make the chart large enough, you can select one of these and remove them all with the Delete key.

Other Chart Types

We have concentrated on XY charts since these are the ones most frequently used by technical people. Most of the techniques we have covered are applicable to other Line, Column and Bar charts. Furthermore the reader now has enough knowledge to be able to work with other chart types (Radar, Area, etc) with some experimentation. We conclude the chapter by looking at a Surface chart, a combination Line/Column chart, and a Bar chart.

Exercise 14: Surface Chart

Excel can make surface plots, that is, a chart from a two-dimensional table. However, this has limitations in that the *x* and *y*-axes are category axes, not value axes. The meaning of this is shown later.

Scenario: The table in Figure 7.17 represents the result of an experiment in which a certain physical quantity was measured as parameters A and B were altered. We wish to show the data graphically.

(a) On Sheet9, enter the values shown in the figure. The text **Parameter B** was typed into A6; the cells A6:A14 were merged and then formatted to have a 90° orientation.

(b) Select B5:G14 and use *Insert | Charts | Other Charts* and select the first Surface chart.

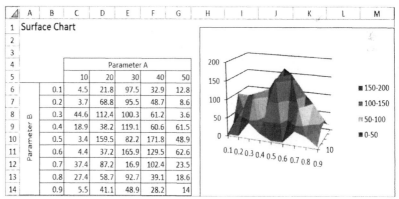

Figure 7.17

(c) With a Surface chart there is no option Format Data Series in the shortcut (right click) menu or in *Chart Tools | Format | Current Selection*. However, the appearance of the data series may be changed with *Chart Tools | Design | Chart Styles*. This feature is available in all chart types, but we opted not to explore it since there ways to achieve the same effects with

XY charts. Note that if the chart is made larger in the vertical direction, the legend expands and the chart has more color bands.

(d) You may wish to experiment with the chart (perhaps changing it to a wire diagram) before you save the workbook.

Exercise 15: Combination Chart

When a chart has more than one data series, it is not necessary to chart them as the same type. When multiple chart types are used in one chart, we have a *combination* chart. We shall make a combination Line and Column chart to demonstrate how mixed types can be visually helpful. The data shown in Figure 7.18 is essentially the same as that used in Exercise 9.

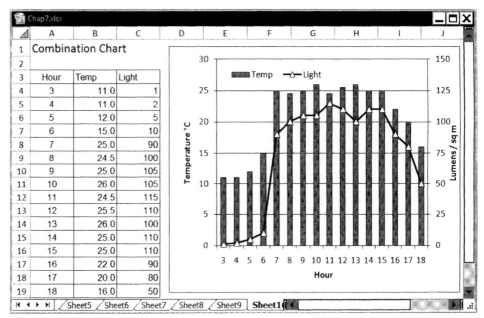

Figure 7.18

We will begin by making a mistake to show a certain feature of Excel.

(a) Having entered all the data, click anywhere within A3:C19 and use *Insert | Charts | Column* and select the first subtype (two columns side by side).

Oh dear! We have three data series. Because the first column has numeric values, Excel treated it as a data series rather than the category series. It used the sequence 1,2,3... for the categories. We could select the Hour series on the chart and remove it with ⌞Delete⌟. The categories (labels on the *x*-axis) would still be incorrect. These

In Excel 2007, the chart feature for using pattern fills was removed (what Microsoft calls *deprecated*).

The functionality was retained for backward compatibility, but the interface to it was excluded. Andy Pope's site (see end of chapter) has an add-in that provides the interface in order to use that functionality.

could be corrected by opening the Select Data dialog in *Chart Tools | Design | Data*. Let's make life easy and use another method.

(b) Delete Hour in A3. Now make the chart with the active cell somewhere within A3:C19. The absence of a header tells Excel to treat the first column as the category data. You can restore the text in A3 once the chart is made.

(c) Right click on one of the Light bars; use *Change Series Chart Type* to select Line.

(d) Follow what was done in Exercise 9 to give the Light series its own axis. This time, with a little planning, we have aligned the tick marks on the two vertical axes; it looks so much better this way when gridlines are present. The range in one scale needs to be a multiple of the other range (30 vs. 150).

(e) Note how the tick marks and labels lie between two data points. This is the default for category charts, but it can be changed on the Format Axis dialog.

(f) Experiment with formatting the Temp data series; the width of the columns and the gaps between the columns are adjustable.

(g) Save the workbook.

Exercise 16: A Bar Chart

If a category chart (any chart type other than XY) gives unexpected results try removing the label at the top of the left-most column used for the chart and remake the chart.

In this Exercise we shall make a bar chart with a difference. The technique we use here can also be applied to make a Gantt chart.

(a) Enter the data shown in Figure 7.19 into Sheet11.

(b) We need to select noncontiguous data. Select A7:A11; hold Ctrl key down while selecting C7:D11; use *Insert | Chart | Bar Chart* and specify the second subtype (two bars in a row). Delete the legend.

(c) Format the first data series (nearest vertical axis) to have no border and no fill, thus making it invisible.

(d) Format the second data series as required.

(e) Use the same technique as in step (f) of Exercise 10 to add text boxes. Save the workbook.

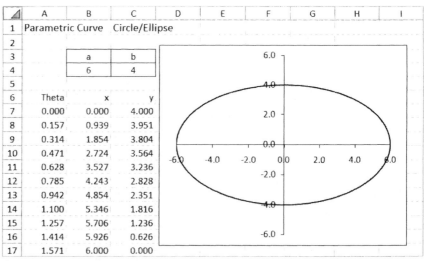

Figure 7.19

Excel has a feature to add Data Labels to charts, but you cannot select another range for them. See the companion website for links to websites that provide free add-ins for adding data labels from other ranges.

Exercise 17: A Parametric Chart

When we want to plot one function, f(x), against another, g(x) we need a *parametric* plot. A simple example would be to plot $R\sin(x)$ against $R\cos(x)$ to generate a circle or radius R. We shall be a little more adventurous and generate a plot that can morph from circle to ellipse using $A\sin(x)$ against $B\cos(x)$.

Figure 7.20

(a) On Sheet 12 of Chap7.xlsx construct a worksheet as in Figure 7.20. Cells B4 and C4 hold the values of A and B, the ratio of which defines the eccentricity of the ellipse. When this ratio is 1, we have a circle. These two cells are named A and B, respectively.

(b) Give the cell A7 the value 0; give A8 the formula =*A7+PI()/20*. Copy this formula down to row 47 to give a final value of 2π.

(c) Give the cell B7 the formula =*a*SIN(A7)* and C7 the formula =*b*COS(A7)*. Use Autofill to copy both down to row 47.

(d) To construct an XY chart, it is necessary to select B7:C47; otherwise Excel will also use column A in its chart. So that you can experiment, make A and B both 6, set the scale of the two axes to ±6, and resize the chart until a circle is seen. You can now experiment with other values of A and B, limiting yourself to values of 6 or less. Save the workbook.

In Problem 3 of this chapter you are asked to generate more complex parametric plots called Lissajous curves.

Exercise 18: Polar Chart

A vector has magnitude and direction; we may denote this using *r* and *θ*. We may also need to refer to its origin. In this exercise we see how to graphically represent a vector with origin 0,0° and end point 15, 60°.

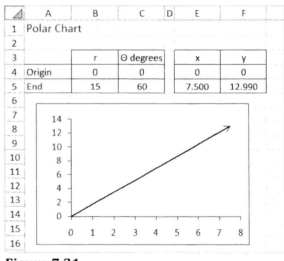

◢	A	B	C	D	E	F
1	Polar Chart					
2						
3		r	Θ degrees		x	y
4	Origin	0	0		0	0
5	End	15	60		7.500	12.990

Figure 7.21

(a) On Sheet 13, enter the text shown in Figure 7.21. Enter the values in B4:C5.

(b) The *x* components are given by *r*cos(*θ*), but we must use angles in radians not degrees. So, for E4 we use =*B4*COS(RADIANS(C4))*. Similarly for the *y* value in F4 use =*B4*SIN(RADIANS(C4))*. Copy these down one row.

(c) When you make the XY chart, Excel will disappoint you: it will treat the data as two data series in rows. Right click the chart, use Select Data, and on the dialog use *Switch Row/Column*.

(d) The arrow is added with *Insert | Shapes*. Make sure it does not extend past the end point of the vector. The magnification tool (lower right corner of the status bar) is useful here. If the chart is selected when you add the shape, then the arrow becomes part of the chart; try moving the chart and the arrow should stay on the line.

Dynamic Charts

We have seen a way of extending the data that is used by a chart. When data added to the end of a table is automatically added to the chart we say the chart is *dynamic*. If the data used to make the chart is an Excel Table (as discussed in Chapter 6), then the chart is made dynamic. You may wish to experiment.

Another way to make a chart dynamic is by the use of the OFFSET function. See the file *DynamicChart.xlsx* on the companion website for an example of its use.

Printing a Chart

If an embedded chart (a chart on a worksheet) is selected when the print command is issued, the printout consists of just the chart. It is expanded to fill the page. You may try this using Print Preview. Conversely, there may be times when you wish to print a worksheet without the chart. Activate the chart; use *Chart Tools | Format | Size* and use the *Size* dialog launcher; on the Properties tab remove the check mark from the *Print Object* box.

URLs for Chart Websites

If you have a charting problem, one of the following sites will very likely have an answer for you. The companion website has an up-to- date list of Excel related links.

Jon Peltier: http://peltiertech.com/Excel/ChartsHowTo

Andy Pope: http://www.andypope.info/charts.htm

Tusha Mehta: http://tushar-mehta.com/excel/charts/

Kelly O'Day: http://processtrends.com/index.htm

Stephen Bullen: http://oaltd.co.uk/Excel/Default.htm

Jan Karel Pieterse:
     http://jkp-ads.com/Articles/ChartAnEquation00.htm

Problems

1. *Using the information from Problem 4 in Chapter 4, make a chart similar to that shown in Figure 7.22.

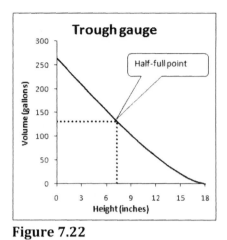

Figure 7.22

2. *The roots $\sin(x) - \cosh(x) + 1$ were found graphically to be exactly 0 and approximately 1.3. Reproduce the chart in Figure 7.23.

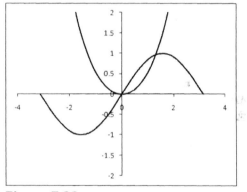

Figure 7.23

3. Show graphically that the equation $x^3 - x + 2 = 0$ has only one real root.

4. Make the chart as shown in Figure 7.24. The technique used in Exercise 10 will not work with a Line chart. The simplest way is to make a column with the same number of points as in the main data series for each line to be added. Other methods include using secondary axes or using error bars. To explore these, see the topic *URLs for Chart Sites* above and search Jon Peltier's site with 'horizontal line'. Add the shape with text, making sure it is part of the chart; it should move with the chart and be present in a copy of the chart.

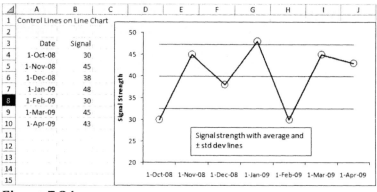

Figure 7.24

5. Refer to Problem 9 in Chapter 2. Make a chart with the number of extractions as the *x*-values and the m_1 values as the *y*-values. Comment on the shape of the curve. What does the curve tell you in practical terms?

6. The left-hand side of Figure 7.25 shows vapor pressure data for three liquids.[1] Thus the vapor pressure of O_2 is 10 mm Hg at 61.55 K. On the right we have the critical temperatures and pressures for the same substances.

[1]J. B. Dence, *Mathematical Techniques in Chemistry*, Wiley, New York, 1975 (page 39).

	A	B	C	D	E	F	G	H	I	J
1				Temperature (K)					Critcal Data	
2		1 mm	10 mm	40 mm	100 mm	400 mm	760 mm		Tc (K)	Pc (mm)
3	O_2		61.55	68.05	73.35	83.35	89.15		153.75	50.1
4	CS_2	198.35	227.45	249.65	267.05	300.15	318.65		551.15	78.0
5	C_2H_4	103.85	118.95	130.85	140.35	158.25	168.45		282.05	50.5

Figure 7.25

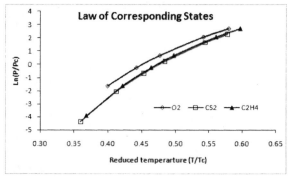

Figure 7.26

According to the Law of Corresponding States, if pressure is expressed as a simple function of the critical pressure, volume as a simple function of the critical volume, and temperature as a simple function of the critical temperature, a general form of the equation of state is obtained which is applicable to all substances. Demonstrate this by constructing the chart shown in Figure 7.26. A careful use of mixed references, together with some copy and paste operations, can save you much time in preparing the data to be plotted.

7.   Prepare a chart similar to that in Figure 7.27. Make up suitable data. The pattern fill using Andy Pope's add-in mentioned in Exercise 15 is optional, but use a format to pick out subcontract work. Note the advice in the sidebar to Exercise 15.

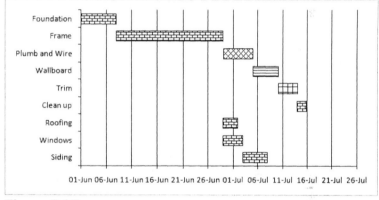

Figure 7.27

8.   Create the chart in Figure 7.28 to demonstrate the addition of two vectors and their resultant. You will need to use the Paste Special method to add the second and third data series.

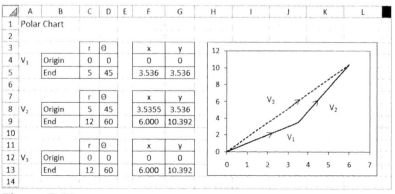

Figure 7.28

[2] D. A. Skoog et al, *Analytical Chemistry*, Saunders College Pub., Fort Worth, 2000 (page 501).

9. In a potentiometric titration, a pair of electrodes (one specific to the ion being measured and the other a reference) is placed in a solution with unknown ion concentration. The voltage of the electrodes is recorded after the addition of each aliquot of titrant. Figure 7.29 shows some typical data.[2] The end point is the volume at the midpoint of the steep rise. Construct the chart with two data series. The differential value is found as $(E_2-E_1)/(V_2-V_1)$ and this value is assigned to the volume at the midpoint of V_1 and V_2. On a separate chart show d^2E/dV^2 vs V.

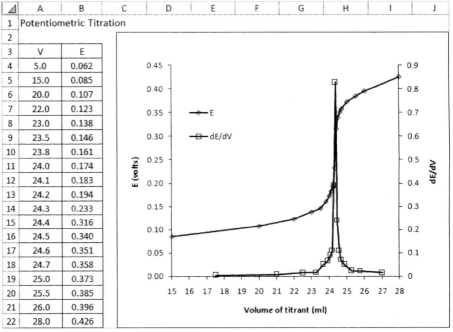

	A	B
1	Potentiometric Titration	
2		
3	V	E
4	5.0	0.062
5	15.0	0.085
6	20.0	0.107
7	22.0	0.123
8	23.0	0.138
9	23.5	0.146
10	23.8	0.161
11	24.0	0.174
12	24.1	0.183
13	24.2	0.194
14	24.3	0.233
15	24.4	0.316
16	24.5	0.340
17	24.6	0.351
18	24.7	0.358
19	25.0	0.373
20	25.5	0.385
21	26.0	0.396
22	28.0	0.426

Figure 7.29

10. Figure 7.30 shows the function y = Abs(Sin(3θ)) plotted on a radar chart with θ going from 0 to 2π in 36 steps.

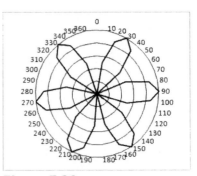

Figure 7.30

Reproduce this chart. Then increase the number of points to get the "rose" diagram with smooth lines.

11. Create the pictograph shown in Figure 7.31 following these steps: (i) make a Line chart, (ii) find a suitable picture, copy it to the worksheet and make it quite small, (iii) copy the reduced picture, (iv) click the chart's data series and use Ctrl +V to copy the picture. This type of chart is clearly not for serious work.

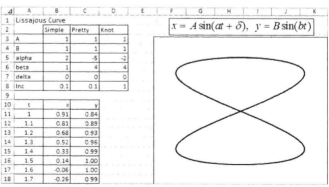

Figure 7.31

12. Lissajous curves result when one sine wave signal is applied to the X plates of an oscilloscope while another is applied to the Y plates. We can achieve this with a parametric plot. Your task is to generate the chart in Figure 7.32. There are 80 data points. The parameters in B3:B7 were used in formulas for the x- and y-values. These can be replaced by values in C3:C7 or D3:C7 to get more complex figures.

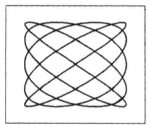

Figure 7.32

Regression Analysis

In this chapter we seek answers to the question: *What equation fits my experimental data?* The general terminology for this type of activity is *regression analysis.* The reader may wish to Google to find how this term came to be used.

We begin by looking at linear functions—functions that can be recast as $y = mx+b$. We explore the use of chart *trendlines* and the Excel functions SLOPE, INTERCEPT, TREND, FORECAST, and LINEST. Then we explore some nonlinear functions, again using trendlines, and the Excel function LINEST and LOGEST. We conclude by showing the use of Excel's Data Analysis tools. In Chapter 12 we will see how Solver can be used for curve-fitting problems, especially for problems where trendlines and Excel function cannot be used.

Least-Squares Fitting

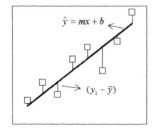

Gauss is credited with developing the fundamentals of the basis for least-squares analysis in 1795 at the age of 18. One speaks about the line of best fit. In this instance, we will restrict ourselves to linear fits. Let the experimental data consist of pairs of x- and y-values. We write the equation of the line of best fit as $\hat{y} = mx + b$, where $\hat{y}$ (read as "y hat") is the predicted value. The vertical displacement between the actual y-value and the predicted $\hat{y}$ for a given x is called the residual. The least-squares criterion requires that we adjust the constants m and b such that the sum of the squares of the residuals, $\Sigma(y_i - \hat{y})^2$ is as small as possible. There are formulas for finding these parameters, but we shall let Excel do the work.

Exercise 1: Trendline, SLOPE, and INTERCEPT

Scenario: A physics student is tasked with finding the thermal coefficient of resistance of a sample. Her experimental results are shown in Figure 8.1.

The textbook told her to work with Equation 1, where R_0 is the resistance at 0°C, R_t is the resistance at temperature t °C and α is the required coefficient.

$$R_t = R_0(1 + \alpha t)$$

Of course, this can also be written as:

$$R_t = \alpha R_0 t + R_0$$

which has the form of the well-known equation of a straight line $y = mx + b$. The slope will be αR_0 and the intercept R_0; hence α can be found from slope/intercept.

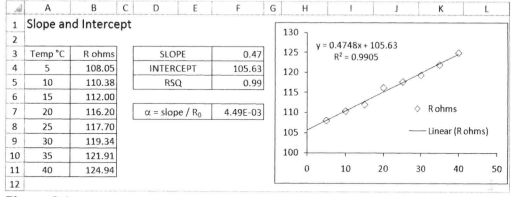

Figure 8.1

(a) Open a new workbook and on Sheet1 enter the text and data shown in columns A and B of Figure 8.1.

(b) Construct an XY chart using the first subtype—markers only.

Now we are ready to add the trendline. We could select the chart and use *Chart Tools | Layout | Trendline | Linear Trendline* to quickly add a trendline. This just adds the trendline; we want more. The same steps, but ending with *More Trendline Options*, will open the required dialog but we shall use the shortcut menu.

R-squared gives a measure of the goodness of the fit. In a sense, it is a measure of how much of the variability in the *y*-values can be accounted for by changes in the *x*-values. Here it is 99%; the rest may be attributed to experimental errors.

(c) Right click a marker on the chart and select *Insert Trendline* from the shortcut menu to open the Trendline dialog (see Figure 8.2).

(d) Clearly, we want a linear trendline, so make that selection. For this demonstration also check the boxes to give us the equation of the best fit and the R-squared value. Our data starts at 5°C but it will be interesting to have the trendline start at 0°C (then it will hit the *y*-axis), so in the *Backwards* box of the *Forecast* group, enter a value of 5. Note that the Trendline Equation box can be dragged around the chart.

We have used the equation $y = mx + b$. Be aware that there are other conventions. In the UK it is $y = mx + c$ and m is called the *gradient*. Statisticians, and the authors of some entries in Excel's Help, like to use $y = a + bx$. So that b may not be the b you are thinking of when you flip through a textbook or glance at Help. Always check what convention is being used.

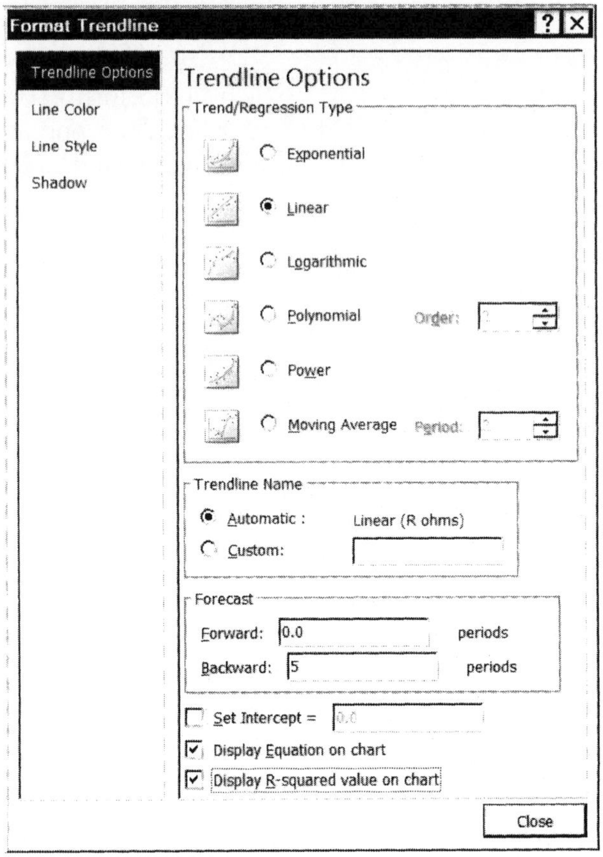

Format Trendline

Trendline Options
Line Color
Line Style
Shadow

Trendline Options

Trend/Regression Type

- ○ Exponential
- ● Linear
- ○ Logarithmic
- ○ Polynomial   Order:
- ○ Power
- ○ Moving Average   Period:

Trendline Name

- ● Automatic :   Linear (R ohms)
- ○ Custom:

Forecast

Forward: 0.0   periods
Backward: 5   periods

☐ Set Intercept = 0.0
☑ Display Equation on chart
☑ Display R-squared value on chart

Close

Figure 8.2

Remember that we want both *slope* and *intercept* in order to compute the coefficient α. The are three good reasons not to just copy the trendline values into cells on the worksheet: (i) We may not use enough significant figures (the Trendline equation can be formatted to show more or less digits), (ii) should we make a correction to the data used to make the plot, we may forget to recopy the Trendline parameters, and (iii) it is an error-prone operation. Rather, we shall get the parameters of the line of best fit using functions.

(e) Enter the text in columns D and E.

(f) The formulas we need in column F are:

    F3:   =SLOPE(B4:B11, A4:A11)
    F4:   =INTERCEPT(B4:B11, A4:A11)
    F5:   =RSQ(B4:B11, A4:A11)
    F6:   =F3/F4 to give α

Note the syntax for the three regression formulas is FUNCTION(*y-value, x-values*). Engineers and scientists generally use *x* before *y* in this context, so be careful.

(g) Save the workbook as Chap8.xlsx.

We see that the trendline values and the function results agree. In the figure the function values are formatted to show only two decimal places in keeping with the experimental data, but you can look at both the trendline equation and the function results to 15 decimals to compare them.

Exercise 2: Interpolation and FORECAST

In the previous exercise we have fitted data for temperatures in the range 5 to 40°C in 5-degree intervals. How would we compute the expected resistance of the sample at (i) 22°C, (ii) 0°C, and (iii) 100°C? The first task is called interpolation (we want a value within the known range); the others are extrapolation (we want a value outside the known range). It is generally safe to interpolate. Extrapolation is risky especially when the value lies far from the know range. Many physical systems appear to behave in a linear fashion over a short range but actually obey a more complex law. A gas obeys the Ideal Gas Law at low pressure and high temperatures but not under other conditions. Of course, the second task is special, we are looking for the intercept—a value we already know!

If you copy a chart from one worksheet to another, the new chart will still be using the data from the original worksheet. But you can copy an entire worksheet by holding [Ctrl] and dragging the tab of the source worksheet to the right (or left) past the next tab.

◢	A	B	C	D	E	F
1	Interpolation, Extrapolation and FORECAST					
2						
3	Temp °C	R ohms		SLOPE		0.47
4	5	108.05		INTERCEPT		105.63
5	10	110.38				
6	15	112.00		Temp	y=mx+b	FORECAST
7	20	116.20		22	116.08	116.08
8	25	117.70		0	105.63	105.63
9	30	119.34		100	153.11	153.11
10	35	121.91				
11	40	124.94				

Figure 8.3

(a) Our completed worksheet will look like Figure 8.3. We can copy much of it from Sheet1. Select from Sheet1 A1:F11, copy it, and paste this to A1 of Sheet2. Use *Home | Editing | Clear* (looks like an eraser) | *Clear All* to remove D5:F7.

(b) Enter the text in D6:F6 and the values in D7:D9.

If we know the parameters for the equation of the straight line, we can find the value of *y* for any *x* with $y = mx + b$. We do this in E7:E9. In F7:F9 we use the FORECAST function to show that if this is our only task we do not need to find the slope and intercept but can have Excel do that "behind the scenes."

(c) Enter these formulas:

E7:   =$F$3*D7+$F$4

You may wish to refer to Help to understand the syntax of the FORECAST function.

F7:   =FORECAST(D7, $B$4:$B$11, $A$4:$A$11).

Note that we have used some absolute references as we wish to copy these formulas.

(d) Copy E7:F7 down to row 9. Remember the double-click on the fill handle trick?

(e) You may wish to return to Sheet1 and extend Trendline to 100°C and compare the chart and the values in E9 and F9.

(f) Save the worksheet.

Exercise 3: The LINEST Function

In this Exercise we use LINEST rather than SLOPE, INTERCEPT, and RSQ to get the parameters for a linear fit. LINEST is more flexible and can give more data, as we shall see in this and subsequent Exercises.

In Figure 8.4 we have the results of a chemistry experiment to measure the enthalpy of solution (ΔH) of ℓ-ascorbic acid at various mole fractions (X).

Temporarily ignore E4:F8 and enter all text and values as shown in Figure 8.4 onto Sheet3. Construct the chart.

(a) LINEST is an array function in that it returns more than one value. With E4:F8 selected, type the formula =LINEST(B4:B20,A4:A20,TRUE,TRUE) using Ctrl + ⇧ Shift + ⏎ to commit it. Note the braces around the formula when viewed in the formula bar.

(b) Save the workbook.

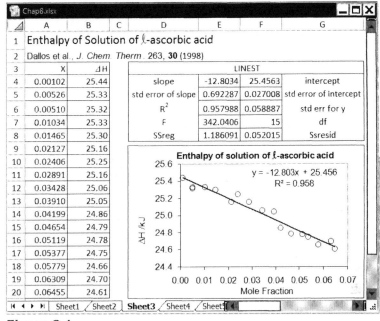

Figure 8.4

When used with four arguments, LINEST returns the slope, intercept, and R^2 value as well as a number of other statistics that we address in a later chapter. You will see that arguments one and two are the y- and x-values as used with SLOPE. When the third argument is TRUE or omitted, LINEST computes the intercept; otherwise it sets the intercept to zero. The statistics in rows below the fit parameter are not returned if the fourth argument is either set to FALSE or is omitted. A two-argument formula such as =LINEST(B4:B20, A4:A20) would just give the slope and intercept.

Exercise 4: Fixed Intercept

Occasionally, you want to get a fit with a fixed intercept. You may, for example, want an intercept of zero or of some other value. If you look at Figure 8.2, there is a setting Set Intercept where you can specify the required intercept value. Getting a zero value with LINEST is simple; you just enter FALSE for argument three. Specifying a value such as 5 needs a "workaround." The lines $y = 1.5x+5$ and $z = 1.5x$ are parallel. For a given x, the y-value equals the z-value plus 5. So if we subtract 5 from each y-value, we get the z line and its intercept is 0. Let's see how we implement that in Excel.

(a) On Sheet4, we will make a worksheet like Figure 8.5. Enter the values and text in columns A, B, and C, and the text in columns E and F.

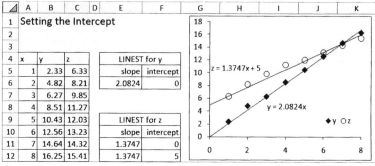

Figure 8.5

(b) Create the chart. Add the trendlines giving the *y*-line an intercept of 0 and the *z*-line and intercept of 5. Delete the trendline entries in the legend. Edit the second trendline equation to show *z*= ... rather than *y*= ...

(c) In E6:F6 the LINEST formula for the *y*-line is
=LINEST(B5:B12,A5:A12,FALSE).

(d) In E11:F11 we want to use *y*-values that have 5 subtracted from them using =LINEST(C5:C12-5,A5:A12,FALSE).

(e) But the 0 in F11 is misleading. In E11 use
=INDEX(LINEST(C5:C12-5,A5:A12,FALSE),1,1) which extracts from the LINEST array the item in row 1, column 1. Since we want the very first element, we could use LINEST(C5:C12-5,A5) but the longer formula shows how any element can be extracted. In F12 enter the value 5 since this is the fixed value.

(f) Save the workbook.

Exercise 5: A Polynomial Fit

Looking at Figure 8.2, one can see that Excel can do more than just linear trendlines. How about LINEST? Can that cope with other than linear functions? We will look at a polynomial fit.

Scenario: An engineer has measured the temperature of an extruder machine's die at various settings of the screw revolution speed. The results are shown in Figure 8.6. He would like an empirical formula to summarize the data.

(a) On Sheet5, enter the text and values shown in rows 1 through 6 of Figure 8.6. If you enter formatted text in two cells (B6:C6), select the range and drag the fill handle; Excel will automatically complete the rest of the test.

(b) Make an XY chart with only markers. Experiment with trendlines with polynomials of order 2 (quadratic), 3 (cubic), and so on. Stop when all the points are fairly close to the line; in this case a fourth order (quartic) gives a reasonable fit. Remember that with six data points, a fifth-order function will give a perfect fit. There is, however, no justification in using this, as the leading coefficient is very small.

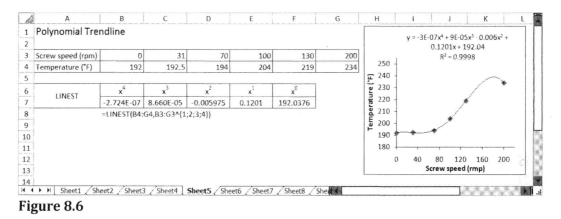

Figure 8.6

At the time of writing, Excel 2007 had a bug that caused it to occasionally drop the leading coefficient in the trendline polynomial equation. Microsoft expects this will be fixed in service pack 2. Make sure you have downloaded it.

Now we will have LINEST generate the same coefficients. The naive way is to insert rows with values of the screw speed raised to the powers of 2, 3, and 4. But there is a better way.

(c) To hold the coefficients of x^4, x^3, x^2, x, and b (or x^0), we need five cells. Select B7:F7 and enter the array formula =LINEST(B4:G4,B3:G3^{1;2;3;4}). Commit it with [Ctrl]+[⇧ Shift]+[↵].

(d) Save the worksheet.

NOTE: In the formula used in step (c) we have an array of constants: {1;2;3;4}, the elements of which are separated by semicolons. This is because our x-values are in columns. Had we made a vertical table, the array would have used commas to separate elements: {1,2,3,4}.

Exercise 6: A Logarithmic Fit (LOGEST)

A simple model for the growth of bacteria predicts that if the initial population is N_0, the population N_t at time t will be given by the following equation, in which B is the reproduction rate.

$$N_t = N_0 \exp(Bt)$$

We can linearize (which means to give it the form $y = mx + b$) by taking natural logs on both sides:

$$Ln(N_t) = Bt + Ln(N_0)$$

Before computers, the normal practice was to convert equations to linear form since fitting to a straight line is relatively simple.

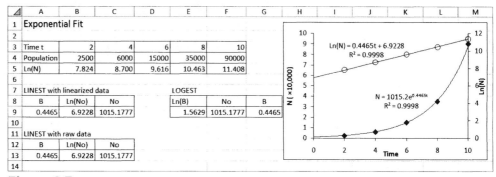

Figure 8.7

For this demonstration, we will find the fitting parameters of some exponential data both with and without linearization.

Figure 8.7 shows the population of a bacteria colony at various times. We wish to estimate N_o and B. For a change of pace, the reader is asked to develop this worksheet without detailed instructions. Row 5 has the linearized data, which we fit with LINEST in row 9. Row 13 also uses LINEST, but we do the logarithm transformation within the formula. Also in row 9 we use LOGEST to do the fit in a most straightforward way. We also check out functions by using a chart with two trendlines whose labels we have edited to remove the default x and y text. The formulas required are:

```
B5:        =LN(B4)
A9:B9:     =LINEST(B5:F5,B3:F3)
C9:        =EXP(B9)
A13:B13:   =LINEST(LN(B4:F4),B3:F3)
C13:       =EXP(B13)
E9:F9:     =LOGEST(B4:F4,B3:F3)
G9:        =LN(E9)
```

The LOGEST function (which can also return fitting statistics) fits our data to $y = km^x$ while the exponential trendline fits it to $y = k\text{Exp}(ax)$. It is left to the reader to show that $a = \text{Ln}(m)$.

As expected, the results of the three methods are in agreement. Remember to save the workbook when done.

The TREND and GROWTH Functions

The parameters in the LINEST and LOGEST arrays can, of course, be used to find the values in the trendlines or to interpolate or extrapolate. However, this is more readily done with TREND (for LINEST fits) and GROWTH (for LOGEST fits). The syntaxes for these functions are:

TREND(known_ y's,known_x's,new_x's,const)
GROWTH(known_y's,known_x's,new_x's,const)

	O	P	Q	R	S	T	U
1	Screw speed (rpm)	0	31	70	100	130	200
2	Temperature (°F)	192	192.5	194	204	219	234
3	TREND	186	193	202	209	215	231

	O	P	Q	R	S	T
1	Time t	2	4	6	8	10
2	Population N	2500	6000	15000	35000	90000
3	GROWTH	2480	6057	14794	36137	88268

Figure 8.8

Figure 8.8 shows the use of TREND to compute the fitted values in the problem posed in Exercise 6 and the use of GROWTH for the problem in Exercise 7. The formulas, which refer to different worksheets, are:

P3:U3:   =TREND(P2:U2,P1:U1) (in the top diagram)
P3:T3   =GROWTH(P2:T2,P1:T1) (the lower diagram)

The two, of course, must be entered as array formulas.

Residuals

Computing and plotting residuals can sometimes reveal otherwise hidden information.

Recall that we have defined *residuals* as the difference between the actual and the predicted values in a curve-fitting problem. If the residuals are the result of normal experimental errors, we would expect them to be distributed randomly above and below the $x = 0$ line. If, on the other hand, the residuals display an observable trend, then one should question the fit.

In the example shown in Figure 8.9 (albeit a very contrived set of data), a linear fit seems very appropriate, but the residuals appear to follow a parabolic rule. Two trendlines were added to the original chart; both give R^2 as 1, but the results from REQ (linear fit) and LINEST (using =INDEX(LINEST(y, x^{1,2}, TRUE, TRUE),3,1) for a quadratic fit) show very slightly different values. The meaningfulness of the x^2 coefficient (called by some *a lurking variable*) will depend very much on the circumstance of the experiment.

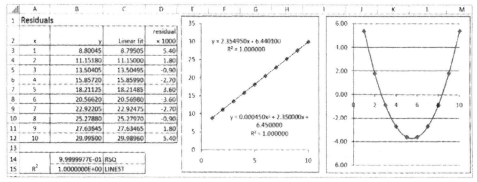

Figure 8.9

Exercise 7: Slope and Tangent

In this Exercise we see how to compute the slope of a polynomial and how to display a tangent line on a chart. Suppose we find the slope m at a point x_0, y_0, then the tangent is the line that obeys $y_0 = mx_0 + b$. Hence, $b = y_0 - mx_0$ and we can find a second point on the tangent using $y = m(x - x_0) + y_0$.

The next table gives the formulas needed to compute approximations to the first and second derivatives of tabulated data. The central formula is generally more accurate but requires that we have a point each side of the point of interest.

Order	Forward	Backward	Central
First	$\dfrac{dy}{dx} = \dfrac{y_1 - y_0}{h}$	$\dfrac{dy}{dx} = \dfrac{y_0 - y_{-1}}{h}$	$\dfrac{dy}{dx} = \dfrac{y_1 - y_{-1}}{2h}$
Second	$\dfrac{d^2y}{dx^2} = \dfrac{y_2 - 2y_1 + y_0}{h^2}$	$\dfrac{d^2y}{dx^2} = \dfrac{y_0 - 2y_{-1} + y_{-2}}{h^2}$	$\dfrac{d^2y}{dx^2} = \dfrac{y_1 - 2y_0 + y_{-1}}{h^2}$

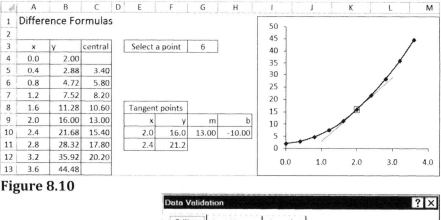

Figure 8.10

Figure 8.11

(a) On Sheet8 of Chap8.xlsx, enter all the text shown in Figure 8.10. Enter the values shown in A3:B13.

(b) The formula to compute the slope using the Central Difference method in C5 is =(B6-B4)/(2*(A5-A4)), and this must be copied down to row 12.

(c) Enter a number in the range 2 to 9 in G3. We will use this as an index to the *x*-values. With the central difference method we cannot use points 1 or 10, so we need to prevent users entering invalid data here. Select G3 and use *Data | Data Tools | Data Validation* to open and complete the dialog box shown in Figure 8.11.

(d) Cells E10 and F10 hold our x_0, y_0 data pair; this is the point on the curve where we want the tangent. Cells E11 and F11 hold the second point on the tangent. In G10:H10 we compute the slope and intercept values of the tangent line.
E10: =INDEX($A$4:$A$13,$G$3)

F10: =INDEX($B$4:$B$13,$G$3)
G10: =INDEX($C$4:$C$13,$G$3)
H10: =F10-E10*G10
E11: =INDEX($A$4:$A$13,$G$3+1)
F11: =G10*E11+H10

(e) Make an XY chart with the data in A3:B13. Using the Copy-&-Paste Special method from Exercise 11 in Chapter 7, add the points defined by E10:F11 as a second series. We want only the first point to be visible, so click on the second point twice and format it to have no marker line or fill.

(f) Add a linear trendline to the new series with appropriate forward and backward projections to make the tangent line visible.

(g) Save the workbook.

Exercise 8: The Analysis Toolpak

Excel has a feature called the Analysis Toolpak, which has a variety of tools that enable the user to generate results without using formulas and formatting. In this exercise we will see the use of the Regression Tool by repeating the problem set out in Exercise 3 for comparison purposes.

(a) Copy A1:B20 from Sheet3 to A1 in Sheet9.

(b) Use the command *Data | Analysis | Data Analysis* and from the resulting dialog select *Regression,* which opens the dialog shown in Figure 8.12.

(c) The *x* range is B3:B20, and the *y* range is A3:A20. Ensure you have checked the Labels box. A suitable output range for our purposes is E5, but you will note that you could output to a new worksheet or workbook. Check the box *Line Fit Plots* to generate a chart. Click the OK button.

If you compare the results in F21 and F22, shown in Figure 8.13, you will see that the slope and intercept are the same as were generated with LINEST in Exercise 3. You will also see that the statistics are in agreement. None of this is surprising as the Tool uses the LINEST function.

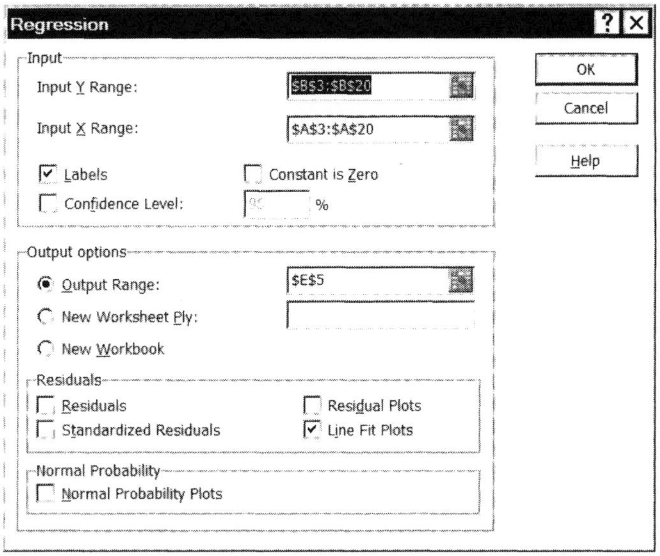

Figure 8.12

	E	F	G	H	I	J	K
5	SUMMARY OUTPUT						
6							
7	*Regression Statistics*						
8	Multiple R	0.9787686					
9	R Square	0.95798797					
10	Adjusted R Square	0.95518717					
11	Standard Error	0.05888709					
12	Observations	17					
13							
14	ANOVA						
15		*df*	*SS*	*MS*	*F*	*Significance F*	
16	Regression	1	1.186090546	1.186091	342.0406	9.7762E-17	
17	Residual	15	0.052015336	0.003468			
18	Total	16	1.238105882				
19							
20		*Coefficients*	*Standard Error*	*t Stat*	*P-value*	*Lower 95%*	*Upper 95%*
21	Intercept	25.4563033	0.027008461	942.5307	1.26E-37	25.3987361	25.5138705
22	X	-12.8033893	0.692286884	-18.4943	9.78E-12	-14.278964	-11.327815

Figure 8.13

There are two major drawbacks to using this Tool. The user has no control over the positioning of the various resulting values and, like all Data Analysis Tools, the results are static. This means that if you make a change in the input data you must remember to rerun the Tool.

Art the time of writing, the Regression Tool has a small bug in that it produces a Column chart when an XY chart is required. The user should right click and change the chart type.

Problems

[1]www.astro.indiana.edu/
catyp/activities/hubble.doc

1. *What function best fits the data[1] in the following table?

Galaxy	Distance (Megaparsec)	Radial Velocity (km/s)
Virgo	15	1200
Perseus	71	5400
Coma	83	6600
Hercules	150	10500
Ursa Major I	313	15600
Leo	337	19500
Corona Borealis	347	21600
Gemini	402	23400
Bootes	650	39300
Ursa Major II	653	40200
Hydra	831	60600

2. *A chemical engineer is studying the rate at which compound X reacts under certain conditions. The following table gives the percentages of X remaining after measured times. Fit the data to $(1-X) = Exp(-kt)$ to determine k using (a) a graphical method, and (b) a single cell with a LINEST or LOGEST formula.

t (sec)	X (percent)
200	0.18
400	0.29
600	0.42
800	0.51

3. *In Problem 3 of Chapter 4 we used numerical differentiation formulas to find di/dt for same tabulated data. Another approach is to use LINEST to get the polynomial coefficients; then from $f(x)$ we can find the coefficients of $f'(x)$. Compare the results from each method.

4. The solution to Problem 9 of Chapter 2 consisted of a table giving the amount of solute m_0 remaining in the water after extraction n.
   (i) Plot this data and add an exponential trendline in the form $m_0 = 5exp(-An)$.
   (ii) Fit the data using the LOGEST function to get parameter B and 5.

(iii) Clearly, 5 results from the fact we started with 5g. How are A and B related to each other and to the data in the experiment?

(iv) Do a mathematical analysis of the experiment to explain the exponential fit.

[2] W. L Friend and A. B. Metzner, *American Institute of Chemical Engineering Journal* **4**, 393, 1958.

5. Fit the data below[2] to the equation $N = aP^k$ by: (i) making a plot and adding a power trendline; (ii) plotting $Ln(N)$ against $Ln(P)$ and adding a linear trendline; and using LINEST. Ensure you understand the relationship between the various fitting parameters. Note that you can plot P vs. N and give both axes a logarithmic scale to get a straight line, but this does not help with regression analysis.

P	N	P	N	P	N
0.46	24.80	10.00	84.50	55.00	195.00
0.53	26.50	17.70	115.00	58.50	193.00
0.63	28.50	18.60	115.00	70.30	189.00
0.74	30.00	25.30	150.00	93.00	245.00
3.00	58.40	31.60	127.00	95.00	245.00
4.20	60.30	32.00	140.00	185.00	315.00
5.00	70.70	37.00	165.00	340.00	380.00
5.60	69.00	41.00	170.00	590.00	480.00

6. A chemist measured the partial pressure of a decomposing gas at various times; see the following table. Make an appropriate chart to show that this data follows the equation $Ln(p_0 / p) = kt$ where p_0 is the pressure at time $t = 0$. What value of k is reported by the trendline. Can you get the same result with a LINEST formula?

t	0	600	1200	1800	2400	3000	3600
P	350	247	185	140	105	78	58

7. *The heat of vaporization of a liquid (ΔH_v) may be found by measuring the liquid's vapor pressure at various temperatures and applying the Clausius-Clapeyron equation, which chemists like to write as:

$$\ln\left(\frac{P_2}{P_1}\right) = \frac{\Delta H_v}{R}\left(\frac{1}{T_2} - \frac{1}{T_1}\right)$$

[3] H. F. Stimson, *Journal of Research of the National Bureau of Standards*, 73A, 493, 1969.

A plot of $1/T$ against $1/P$ where T is measured in Kelvin and P in torr will give a straight line with a slope $-\Delta H_v/R$ where R has the value 8.3145 J mole^{-1} K^{-1}. From the following data[3] find ΔH_v for water.

T (K)	313	323	333	343	353
P (torr)	55.364	92.592	149.51	233.847	355.343

8. Make a plot of the following data and add two trendlines, one quadratic and the other cubic. Format the cubic trendline as a dotted line. Hint: Remember the *Selection* group in *Chart Tools | Format.*

x	573	534	495	451	395	337	253
y	1000	800	600	450	300	200	100

9. *A sociological study in 1976 tested the hypothesis that the larger the city the more rushed were the inhabitants. Google with Pace of Life for more details. The table that follows lists some results. Which model best fits the data; (i) a power model $V=kP^a$, or (ii) a logarithmic model $V=m\text{Ln}(P)+c$?

Location	Population	V (ft/sec)
Brno, Czechoslovakia	341,948	4.81
Prague, Czechoslovakia	1,092,759	5.88
Corte, Corsica	5,491	3.31
Bastia, France	49,375	4.90
Munich, Germany	1,340,000	5.62
Psychro, Crete	365	2.76
Itea, Greece	2,500	2.27
Iraklion, Greece	78,200	3.85
Athens, Greece	867,023	5.21
Safed, Israel	14,000	3.70
Dimona, Israel	23,700	3.27
Netanya, Israel	70,700	4.31
Jerusalem, Israel	304,500	4.42
New Haven, U.S.A.	138,000	4.39
Brooklyn, U.S.A.	2,602,000	5.05

[4] F. E. Croxton et al., *Applied General Statistics*, Prentice-Hall, Englewood Cliffs, N J, 1967 (page 390).

10. *The following data[4] records observation of the number of chirps per 20 seconds of crickets as a function of temperature. What relationship do you find? A search of the Internet found two comments that if you count the chirps in

15 seconds and add 40 (one said 37) you get a good estimate of the temperature in °F. Does this data agree with these comments?

T (°F)	46	49	51	52	54	56	57	58	59	60
chirps	40	50	55	63	72	70	77	73	90	93
T (°F)	61	62	63	64	66	67	68	71	71	72
chirps	96	88	99	110	113	120	127	137	137	132

11. The table that follows shows the results of an enzyme kinetics experiment. The quantity V is the velocity of the reaction, while $[S]$ is the concentration of the substrate S. Ideally, this data should be fitted to the Michaelis-Menten equation to find K. Traditionally, biochemists linearize the M-M equation to give the Lineweaver-Burke equation and then plot $1/V$ against $1/[S]$. What value of K is obtained using a trendline and using LINEST? We revisit this problem in Chapter 12 and use Solver to make a direct fit.

$$\text{Michaelis-Menten Eqn: } V = \frac{V_{max}[S]}{[S]+K}$$

$$\text{Lineweaver-Burke Eqn: } \frac{1}{V} = \frac{K}{V_{max}}\frac{1}{[S]} + \frac{1}{V_{max}}$$

[S] (mM)	V (mM/sec)
8.33	3.62E-06
5.55	3.39E-06
2.77	2.75E-06
1.38	1.99E-06
0.83	1.49E-06

VBA User-Defined Functions

User-defined functions (UDF) are also called *custom functions* by some authors.

Visual Basic for Applications (VBA) is an important part of Microsoft Office. When used within Excel, it enables you to write modules that may be subroutines or functions. A subroutine performs a process; we look at these in the next chapter. A function returns a value to a cell (or a range) in the same way as a worksheet function. Collectively, subroutines and functions are called *modules* or *macros*. To add some confusion, the word *module* is also used for the place where one codes one or more macros.

If you have experience with any programming language, you will be familiar with many of the topics covered in this chapter. If you are not yet a programmer, VBA is a great way to begin. The emphasis in this chapter is on coding, so we will use simple examples. Later chapters make use of this skill to code more useful functions.

Why and when do we use user-defined functions (UDF)? Just as it is more convenient to use =SUM(A1:A20) rather than =A1+A2+...+A20, a user-defined function may be more convenient when we repeatedly need to perform a certain type of calculation for which Microsoft Excel has no built-in function. Once a user-defined function has been correctly coded, it may be used in the same way as a built-in worksheet function.

Before you write a user-defined function, make sure that it is not already provided by Excel. The built-in functions are more efficient than user-defined functions. After you have written a function, you must test it thoroughly with a wide range of input values.

Security Note

While macros are indispensable, they are also a source of danger. A macro (primarily subroutines) may contain malicious code. Office 2007 incorporates various security features, but it is the user's responsibility to protect his work. If you unexpectedly receive a file, do not open it even if it appears to come from a friend or colleague. Check with the sender first.

Macro-enabled files have the extension **xlsm**.

One of the new security features results in Excel files containing macros being given a different extension. When the newly created file is saved, we need to specify that it is a *macro-enabled* file. The file is then saved as, for example, Chap9.xlsm. Note the *m* at the end of the extension.

You may need to add the Developer tab to the Ribbon.

You may need to adjust the security setting before you can work with modules. First you need to add the Developer tab to your Ribbon. Click the Office icon; the *Excel Options* is found at the bottom of the Office dialog. In the *Popular* tab fairly near the top, check the box *Show Developer tab in the Ribbon* and return to the worksheet. Next open the *Developer tab* and in the *Code* group click the *Macro Security* tool to open a dialog shown in Figure 9.1. If you choose *Disable all Macros with Notification*, you will be presented with a security question each time you open a file containing a macro.

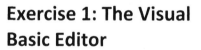

Figure 9.1

Exercise 1: The Visual Basic Editor

The editor is where we do the coding. Before we look at the Visual Basic Editor (VBE), it will be helpful to have an Excel file ready. Then we can explore the VBE.

Use the *Save As* command on the Office tab to display a tool for saving as a macro-enabled file. Or select this type in the *Save As Type* box of the *Save* dialog.

(a) Open Excel and save the new blank workbook as a macro-enabled file named Chap9.xlsm. Use *Developer | Coding | Visual Basic* or the shortcut $\boxed{\text{Alt}} + \boxed{\text{F11}}$ to open the VBE; see Figure 9.2. The top part of the VBE window displays a menu and a toolbar. To the left we have the *Project* window. The top part of the right-hand side is the *Module* window (yours will be empty at this point), and the lower part is the *Immediate* window.

(b) If the Immediate window is not visible, use the menu command *View | Immediate Window*.

We are just scratching the surface of VBA, so we do not have time to explore the VBE window in depth.

(c) Move the mouse pointer into the Immediate window and click, or use the shortcut $\boxed{\text{Ctrl}}$+G. Type ?3*4 and tap $\boxed{\leftarrow}$. The result 12 is displayed. Hence the name Immediate; we use this area mainly for testing short pieces of code or issuing brief VBA commands.

(d) Your module window is most likely empty, so we will open a new module. Ensure that one of the items in the Chap9.xlsm project is selected. Use the menu command *Insert | Module*. Note how Module 1 is added to the project tree.

(e) In preparation for the next exercise, in the module window type Function TriArea() and hit $\boxed{\leftarrow}$. Note how VBE helpfully adds End Function.

(f) Return to Excel: you can click the appropriate item in the Windows taskbar or the Excel icon on the VBE toolbar.

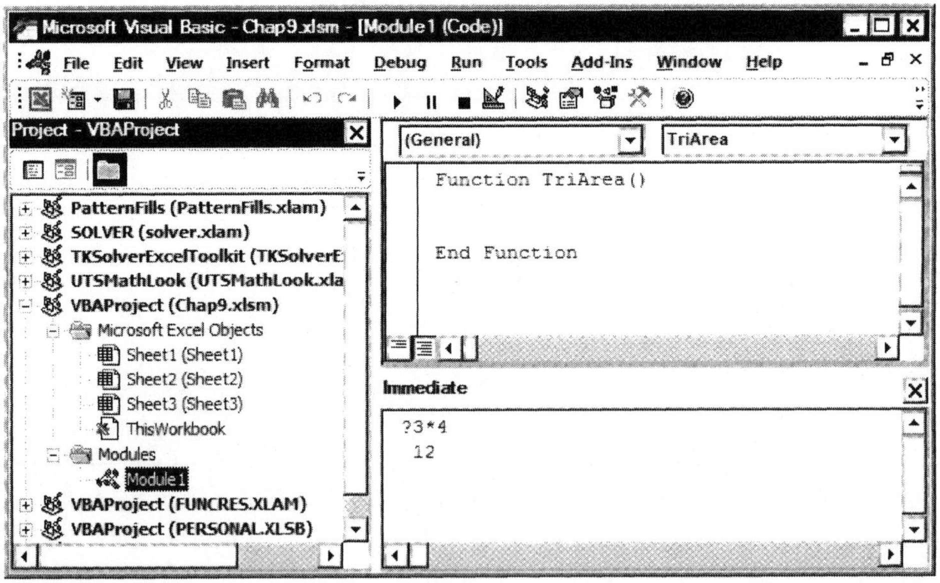

Figure 9.2

Syntax of a Function

To successfully code a function, you need two skills. The first is the ability to compose, in English and mathematical symbols, the set of rules that will yield the desired result. This is called the *algorithm*. The second is the ability to translate the algorithm into the Visual Basic language. Like all languages, both natural and computer, Visual Basic has a set of rules known as the language syntax. Figure 9.3 outlines the syntax for a user-defined function.

For simplicity, the options item *[Private / Public]* has been omitted from before the word Function.

The *name* used for a function must not be a valid cell reference such as AB2, nor may it be the same as the name assigned to a cell or a region. If you make this mistake, the cell that calls your misnamed function will display #REF!

Function name [(arglist)][**As** type]
    [statements]
    [name = expression]
    [Exit Function]
    [statements]
    [name = expression]
End Function

name	The name you wish to give to the function.
arglist	List of arguments passed to the function. Arguments are separated from each other by commas.
type	The data type of the value returned by the function.
statement	A valid Visual Basic statement.
expression	An expression to set the value to be returned by the function.

Items shown within square brackets [...] are optional.
Words in bold must be typed as shown (reserved words).
Each statement must begin on a new line. If a statement is too long for one line, type a space followed by an underscore character and complete the statement on the next line. Do not split a word using this method.

Figure 9.3

Exercise 2: A Simple Function

In this Exercise we write a user-defined function to calculate the area of a triangle given the length of two sides and the included angle: $Area = \frac{1}{2}ab\mathrm{Sin}(\theta)$. A worksheet formula is also used to confirm the VBA result.

If you edit a UDF that is already used in a worksheet, then the worksheet must be recalculated (using F9) to have the function report its new value.

◢	A	B	C	D	E
1	Test function to compute area of triangle				
2					
3	SideA	SideB	Angle	Formula	Function
4	1	2	90	1	1
5	2	2	45	1.414214	1.414214
6	2	2	60	1.732051	1.732051

Figure 9.4

(a) Open Chap9.xlsm and on Sheet1 type the entries shown in A1:E3 and A4:C6 of Figure 9.4. The formula in D4 is =0.5 * A4 * B4 * SIN(RADIANS (C4)) and computes the area so that we may test our function. Copy this down to row 6. Leave E4:E6 empty for now.

(b) Use Alt+F11 to open the VBE window. Click on Module1 in the Project window. The window title should read Chap9.xlsm - [Module1 (Code)]. One of the most common errors for VBA beginners is entering the code in the wrong place. For our purposes, the only correct place is on a general module, not a worksheet or workbook module.

(c) Enter this code exactly as shown using ↵ at line ends and Tab⇆ to indent:

```
'Computes the area of a triangle given
'top sides and included angle in degrees
Function TriArea(SideA, SideB, Theta)
    Alpha = WorksheetFunction.Radians(Theta)
    TriArea = 0.5 * SideA * SideB * Sin(Alpha)
End Function
```

(d) Return to the worksheet and in E4 enter the formula =TriArea(A4, B4, C4). Note that as you type =Tri, your function appears in the popup window in the same way that worksheet functions do. Copy the formula down to row 6. The values in the D and E columns should agree. If they differ, return to the module sheet and correct the function. Remember to press F9 to recalculate the worksheet after editing a function.

(e) Save the Chap9.xlsm file.

Although this is a rather simple function, it demonstrates some important Visual Basic features. We now examine each line of the TRIAREA function.

1   This is a comment; the initial single quote (apostrophe) ensures this. A statement may end with a comment: for example, *x= srt(b) 'find the square root of x.*
2   Another comment.
3   The *Function* is displayed in blue in the VBE to indicate a keyword. We chose the name TriArea; a function name can be anything but a keyword. Our function has three arguments. Arguments are passed from the formula in the worksheet to

the function heading by their position, not by their names.

4   This is an *assignment statement*: we give a value to the variable *Alpha*. We do so using an Excel function, so we need to use the WorksheetFunction before Radians. The complete syntax for referencing a worksheet function is *Application.WorksheetFunction.FunctionName*, but the first word may be omitted as it is in our function. You may also use the syntax *Application.FunctionName.*

Did you notice that when you had typed WorksheetFunction, VBE offered a choice of functions? We explore this later. Did you also notice that you could have typed a lowercase *f*? VBE would fix that when you pressed ⏎.

5   This is another assignment statement. There must be at least one statement that assigns a value to the function. In this statement we use the VBA sine function. We might have typed VBA.Sin(Alpha), but this would be redundant.

6   The *End Function* statement is required as the final line.

Naming Functions and Variables

Try to use short but meaningful names for variables, functions and arguments. These three simple rules must be followed:

(i)   The first character must be a letter. Visual Basic ignores uppercase and lowercase. If you use the name term in one place and Term elsewhere, Visual Basic will change the name to match the last used form.

(ii)   A name may not contain a space, a period (.), exclamation point (!), @, $ or #.

(iii) A name may not be a VBA restricted keyword. A full list of reserved keywords is hard to find. However, it is not necessary to know them because, if you try to use one, VBA highlights the word and displays an error message. Generally this will read "Identifier expected," but certain keywords generate other messages. Note that, generally, VBA displays keywords in blue.

Some cautionary notes on naming variables, functions and modules are in order. If you avoid dictionary words like *Range,* you are less likely to run into conflicts with keywords. Variable names such as *myRange* are safe. If you name a function *Extract,* you get no warning until you run it and then the message is terse: That function is invalid. Using *View | Properties Window,* you can name a module something other than *Module3.* Use "odd" names. The author once named a module *Pi,* which caused every UDF in the workbook that used that name for a variable to report an error.

Worksheet and VBA Functions

The mathematical functions available within VBA are shown in Figure 9.5. Details of other functions may be found by searching for String functions or Date functions in the VBA Help.

You cannot use a worksheet function when VBA provides the equivalent function even when the name is not the same. So none of the worksheet trigonometric functions SIN, COS, or TAN may be used, but ASIN and ACOS are permitted. The worksheet function SQRT cannot be used since VBA includes the equivalent SQR function. You may, however, use the worksheet function MOD because Mod in VBA is an operator, not a function. Use the Help facility in the VB Editor to see a list of which worksheet functions are available for use in VBA code.

Abs(x) The absolute value of x.

Atn(x) Inverse tangent of x. Other inverse functions may be computed using trigonometric identities such as: $Arcsin(X) = Atn(X / Sqr(-X * X + 1))$. For more information search Visual Basic Help for Derived math functions.

Cos(x) The cosine of x, where x is expressed in radians.

Exp(x) The value e^x.

Fix(x) Returns the integer portion of x. If x is negative, Fix returns the first negative integer greater than or equal to x; for example, Fix(-7.3) returns -7. See also Int.

Int(x) Returns the integer portion of x. If x is negative, Int returns the first negative integer less than or equal to x; for example, Int(-7.3) returns -8. See also Fix.

Log(x) The value of the natural logarithm of x. Note how this differs from the worksheet function with the same name which, without a second argument, returns the logarithm to base 10. In VBA, the logarithm of x to base n may be found using the statement y = Log(x)/Log(n).

Mod In Visual Basic this is an operator, not a function, but it is similar to the worksheet MOD function. It is used in the form number Mod divisor and returns the remainder of number divided by divisor after rounding floating-point values to integers. The worksheet function and the VBA operator return different values when the number and divisor have opposite signs; see Help for details.

Rnd(x) Returns a random number between 0 and 1.

Sgn(x) Returns -1, 0, or 1 depending on whether x has a negative, zero, or positive value.

Sin(x) The sine of x, where x is expressed in radians.

Sqr(x) Square root of x.

Tan(x) The tangent of x.

Figure 9.5

The method VBA uses is called *Banker's Rounding*, but there is no evidence of bankers using this method!

Exercise 3: When Things Go Wrong

If you try to code a user-defined function that references an unavailable function (e.g. WorksheetFunction.Sin(Alpha)), the worksheet cell in which your user-defined function is called will display #VALUE!

The VBA Round function differs from the Excel ROUND: When the last digit to be rounded is 5, an even number is always produced. Round(4.5,1) gives 4, while Round(5.5,1) gives 6.

Recalling the adage "To err is human, but to really mess up you need a computer," we will make an error in a module and lock our worksheet. Do not worry, we can fix the problem. It is very likely that you will accidentally make such an error, so it is good to know what is needed to correct it.

(a) Open the VBE and change line 5 of TriArea by replacing the equal sign by a minus sign. Do not press [Enter ↵] and do not use *Debug | Compile*. Just return to the worksheet.

(b) Press [F9] to recalculate the worksheet. Excel returns you to the VBA Editor window and displays an error dialog box. Click the OK button. Note that the function header is highlighted in yellow. Correct the error by replacing the minus sign by an equal sign. The yellow highlighting does not disappear.

(c) Return to the worksheet. Try to do something like changing the active cell. Nothing works; the worksheet (indeed the whole workbook) is locked.

(d) Return to the VBE and use the command *Run | Reset* to remove the highlighting. Now when you go to the worksheet and press [F9] all is well again.

In short, an error in a module can cause a worksheet to lock. The remedy is to use the command *Run | Reset* to reset the module.

This exercise should help you with syntax error. Another type of problem is the logical error. This occurs when you have coded your function correctly as far as Visual Basic is concerned (the syntax is correct), but the wrong answer results from an error in the algorithm. Some techniques for solving this type of problem are explored in a future exercise.

Programming Structures

The normal flow in any computer program (and our function is a small computer program) is from line to line. This is called a sequential structure. In Exercise 2 line 4 is executed, followed by line 5, and so on. Anything that changes this flow is called a *control structure*. In the next exercise we look at a branching or decision structure. This structure gives the program two or more alternative paths to follow depending on the value of a variable. The other control structure is the repetition or looping structure, which we explore in later exercises. The code within a loop is executed one or more times.

Exercise 4: The IF Structure

In Chapter 5 we looked at conditional formulas using the IF function to make a selection based on a test. We do the same in a macro using the IF...ELSE structure whose syntax is shown in Figure 9.6. The items enclosed in square brackets are optional. There is a slightly simpler syntax for a one-line IF statement in which statements are separated by colons. For details on logical expressions (a.k.a conditions) refer back to Chapter 5.

Syntax for IF...ELSE

If condition **Then**	condition	Expression that is True or False.
[statements]	statements	One or more statements
[**ElseIf** condition-n **Then**		executed if condition is True.
[elseifstatements]] ...	condition-n	Numeric or string expression
[**Else**		that evaluates True or False.
[elsestatements]]	elseifstatements	One or more statements executed if associated
End If		condition-n is True.
	elsestatements	Statements executed if no previous condition-n expressions are True.

Note: A *condition* is the same as what we called a *logical expression* in Chapter 5.

Figure 9.6

Exercise 5: Boolean Operators

In this exercise we will use the short and the long form of the IF statement, and show the use of the AND and OR Boolean operators.

We will construct a function that reports what type of triangle is formed when given the length of the three sides. Before coding this, we need to think more about the algorithm. We might make a list of the things we know about triangles and their sides:

(i)   One side is always shorter than the sum of the other two.
(ii)   In an equilateral triangle all the sides are equal.
(iii)   In an isosceles triangle two sides are equal.
(iv)   Pythagoras' theorem is true with a right angle triangle.

How do we know which is the hypotenuse if we are given just three values? How many pairs of sides must be compared to establish that we have an isosceles triangle? These questions are readily answered if the values for the sides are in descending order. The reader may wish to complete the algorithm before proceeding.

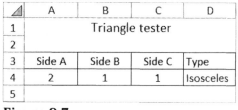

◢	A	B	C	D
1		Triangle tester		
2				
3	Side A	Side B	Side C	Type
4	2	1	1	Isosceles
5				

Figure 9.7

The indentations in the code are used only for readability; they are not required by syntax.

A UDF returns a value to the cell(s) containing the formula that calls it. A UDF cannot change the values in other cells. So the cells referenced by *a*, *b* and *c* do not change in the worksheet. A UDF cannot format a cell.

(a)   On Sheet2 of Chap9.xlsm construct the worksheet shown in Figure 9.7. The cell D4 contains the formula = Tritype(A4, B4, C4). It will return the value #NAME? until we have coded the function.

(b)   Open the VBE, add a second module to the Chap9.xlsm project, and code the following function.

```
Function Tritype(a, b, c)
'Sort the three sides in ascending order
   If b > a Then holder = a: a = b: b = holder
   If c > a Then holder = a: a = c: c = holder
   If c > b Then holder = b: b = c: c = holder
' Determine triangle type
  If a > b + c Then
        Tritype = "None"
    ElseIf a * a = b * b + c * c Then
        Tritype = "Right"
    ElseIf (a = b) And (b = c) Then
        Tritype = "Equilateral"
    ElseIf (a = b) Or (b = c) Then
        Tritype = "Isosceles"
    Else
        Tritype = "Scalene"
```

```
                    End If
                  End Function
```

(c) Return to the worksheet and press [F9]. Experiment with other values for the three sides to test the function.

Make sure you follow the logic in the sorting section. We use a variable called *holder* to temporarily store a value when we need to exchange the values of two other variables.

We made a second module for the user-defined function of this exercise. This was not essential; we could have added it to Module1. However, there is a problem with having more than one function on a single module. If any one of the functions contains an error, then all functions on that module return error values on the worksheet. This can be confusing, especially for beginners. There are other considerations that help you decide whether to put more than one function on a single module, but these relate to the use of Public or Private in the Function header— a topic we will not be exploring.

The SELECT Structure

Visual Basic for Applications provides another branching structure called the SELECT CASE structure. Its syntax is shown in Figure 9.8.

Syntax for SELECT...CASE		
Select Case testexpression 　　[**Case** expressionlist-n 　　　[statements-n]] . . . 　　[**Case Else** 　　　[elsestatements]] **End Select**	testexpression expressionlist-n	Any numeric or string expression. A list of one or more of expression types separated by commas. Valid expression types are: expression, expression To expression, Is comparisonoperator expression.
	statements-n	One or more statements executed if testexpression matches any part of expressionlist-n.
	elsestatements	One or more statements executed if testexpression does not match any of the Case clauses.

Figure 9.8

Examples of an expression include:
(i)　A simple value such as 100;
(ii)　*Smaller-value To larger-value*; as in 0 To 20.

(iii) *Is > some-value*; as in `Is > 20`. Clearly all comparison operators (>, < , >=, <=) are permitted.

When *testexpression* matches one of the expressions, the statements following that Case clause are executed up to the next Case (or End Select) clause. Control then passes to the statement following End Select. When *testexpression* matches more than one expression, only the statements for the first match are executed. The statements following Case Else are executed if no match is found in any of the other Case selections. It is advisable always to use a Case Else statement to handle unexpected testexpression values.

These two functions give the same results:

```
Function Test1(a, b)                    Function Test2(a, b)
  If a > b Then                           Select Case (b - a)
    Test1 = "A is larger"                   Case Is > 0
  ElseIf b > a Then                           Test2 = "B is larger"
    Test1 = "B is larger"                   Case Is < 0
  Else                                        Test2 = "A is larger"
    Test1 = "A & B are equal"             Case Else
  End If                                     Test2 = "A & B are equal"
End Function                              End Select
                                        End Function
```

Exercise 6: Select Example

The number of real roots of the quadratic $ax^2 + bx + c = 0$ is determined by the value of the discriminant $d = b^2 - 4ac$. In this exercise we write a function to return a value indicating the number of real roots for a quadratic equation.

The expression
$d = (b * b) - (4 * a * c)$
could have been coded without the parentheses. But they do make it easier to read the expression because of the way VBE spreads out arithmetic expressions.

(a) Open the VB Editor, insert another module, and enter the function shown here.

```
Function RootCount(a, b, c)
  d = (b * b) - (4 * a * c)
  Select Case d
        Case 0:         RootCount = 1
        Case Is > 0:    RootCount = 2
        Case Else:      RootCount = 0
  End Select
End Function
```

◢	A	B	C	D
1	Number of real root roots of a quadratic			
2				
3	a	b	c	Roots
4	1	3	-15	2
5				

Figure 9.9

(b) Set up Sheet3 as in Figure 9.9 to test the function. The function is called in D4 with =RootCount(A4, B4,C4).

(c) Save the workbook.

The FOR..NEXT Structure

In a looping structure a block of statements is executed repeatedly. When the repetition is to occur a specified number of times, the FOR...NEXT structure is used. The syntax for this structure is given in Figure 9.10.

The reader is strongly advised never to alter the value of the *counter* within the For...Next loop, as the results are unpredictable.

Exercise 7: Example Using FOR...NEXT

For our example we will write a function to find the sum of the squares of the first *n* integers. While we are learning VBA, it is good to use an example where the answer is known. In this case the answer is given by: *(n/6)(n+1)(2n+1)*.

(a) Open the VBE, insert another module, and enter the folowing function.

```
Function SumOfSquares(n)
    SumOfSquares = 0
    For j = 1 To n
        SumOfSquares = SumOfSquares + j ^ 2
    Next j
End Function
```

(b) Set up Sheet4 as in Figure 9.11 to test the function. The cell with value 12 has been named as *N*.

(c) Save the workbook.

Syntax of FOR...NEXT

For counter = first **To** last [**Step** step]
    [statements]
    [**Exit For**]
    [statements]
Next [counter]

counter	A numeric variable used as a loop counter.
first	The initial value of counter.
last	The final value of counter.
step	The amount by which counter is changed each time through the loop.
statements	One or more statements that are executed the specified number of times.

The *step* argument can be either positive or negative. If *step* is not specified, it defaults to one.

After each execution of the statements in the loop, *step* is added to counter. Then it is compared to last. When *step* is positive, the loop continues while *counter* <= end. When *step* is negative, looping continues while *counter* >= end.

The optional *Exit For* statement, which is generally part of an IF statement, provides an alternate exit from the loop.

Figure 9.10

◢	A	B	C
1	Sum of the Squares of first N integers		
2			
3	N	Function	Formula
4	12	650	650
5			
6	Function call	=SumOfSquares(N)	
7	Formula	=(N/6)*(N+1)*(2*N+1)	

Figure 9.11

The Excel Object Model: An Introduction

From a nontechnical point of view, the Excel Object Model is a detailed blueprint of how Excel operates behind the scenes. The model consists of *objects*. Examples of objects are a workbook, a worksheet, a range. A group of objects of the same type is called a *collection*. The *Workbooks* collection contains all the open workbook objects. The term *Workbooks(1)* refers to the first open workbook, while *Workbooks("Data.xlsx")* refers to a workbook by name.

Objects have *methods* and *properties*. Thus we might see subroutine statements with terms such as *Workbook(1).Close* and *Activesheet.Delete* where Close and Delete are methods. An expression such as *Worksheet(1).Name* is a reference to a property of Worksheet(1).

For this chapter, our interest is in ranges A range may be a cell, a row, a column, a selection of cells containing one or more contiguous blocks of cells, or a 3D range.

A range is a strange object. Consider the range A1:C10. It contains other ranges such as A2:B4 and C9, but each of these is itself a range. So the range object is also a range collection; this is the only collection in Excel that does not use a plural name. Surprisingly, there is no *Cell* object. You may see code using a term such as *Cells(1,1)*, but this is just another way of referencing Range ("A1").

When working with Functions, we need to know about two range properties: *Count* and *Value*. Very often the term *Value* is omitted since this is the default property.

Exercise 8: FOR EACH—Resistors Revisited

The *For Each* structure may be used with any collection. This structure references each member of the collection in turn with code such as: *For Each MyCell in MyRange...Next.*

In Exercise 5 of Chapter 2 and again in Exercise 3 of Chapter 5, we found the equivalent resistance of a number of resistors in parallel. For comparison, we will use both *For Next* and *For Each* to find the equivalent (or effective) R value.

(a) Open the VBE, insert another module, and enter the two functions shown here.

```
Function EquivR(myRange)
   recipR = 0
   For i = 1 To myRange.Count
     If myRange(i).Value > 0 Then
        recipR = recipR + (1 / myRange(i).Value)
     End If
   Next
   If recipR > 0 Then
      EquivR = 1 / recipR
   Else
      EquivR = "Error"
   End If
```

```
End Function

Function EffectR(myRange)
    recipR = 0
    For Each myCell In myRange
      If myCell.Value > 0 Then
          recipR = recipR + (1 / myCell.Value)
      End If
    Next
    If recipR > 0 Then
        EffectR = 1 / recipR
    Else
        EffectR = "Error"
    End If
End Function
```

◢	A	B	C	D	E
1	**Resistors in Parallel**				
2					
3	R	1240	1800	2000	0
4					
5		EquivR		EffectR	
6	R$_e$	540 ohms		540 ohms	

Figure 9.12

(b) Set up Sheet5 as in Figure 9.12 to test the function. The Cell B6 has the formula =ROUND(equivR(B3:E3),-1) and uses the Custom Format of *0 "ohms"*. We could have VBA do the rounding with this statement:
EffectR = WorksheetFunction.Round(EffectR, -1)
It is left as an exercise for the reader to experiment with the correct place to enter this code in the macro.

Exercise 9: The DO ... LOOP Structure

Whereas the FOR...NEXT structure is used for a specific number of iterations through the loop, the DO...LOOP structures are used when the number of iterations is not initially known but depends on one or more variables whose values are changed by the iterations. There are two syntaxes for this structure; the Pre-test and the Post-test forms. These are shown in Figure 9.13.

We have the option of looping until a condition becomes true or while a condition is true. In the Pre-test form (Syntax 1), the condition is test *before the first statement* within the loop is executed while in the Post-test (Syntax 2) the test occurs *after the last statement* has been executed. This means that with syntax 2, the statements within the loop are executed at least once regardless of the value of the condition at the start of the O structure.

Syntax 1 for the DO statements

Do {While | Until} condition
    [statements]
    [**Exit Do**]
    [statements]
Loop

        Syntax 2 for the DO statements
Do
    [statements]
    [**Exit Do**]
    [statements]
Loop {While | Until} condition

Figure 9.13

If you inadvertently end up with an infinite loop, your worksheet will "hang." Use `Esc` or `Ctrl`+`Break` to terminate the function.

Typically, the condition in the UNTIL or WHILE phrase refers to one or more variables. The programmer is responsible for ensuring that the variables eventually have values such that the terminating condition is satisfied. The only exception is when a conditional EXIT statement is used to terminate the loop. Should the terminating condition never be reached, you have an *infinite loop*.

In Chapter 10 we will use the Do Loop structure to find the root of an equation using an iterative method. But for the example here, we show its use to compute Exp(*x*) using the Maclaurin series.

$$\exp(x) = \sum_{k=o}^{\infty} \frac{x^k}{k!} = 1 + x + \frac{x^2}{2!} + \frac{x^3}{3!} \dots$$

This is known to be a convergent series. Also we observe that:

$$term_k = \frac{x^k}{k!} \quad \text{and} \quad term_{k+1} = \frac{x^{k+1}}{(k+1)!}$$

$$\therefore term_{k+1} = term_k \times \frac{x}{k}$$

We will make use of this recursive relationship. But how many terms shall we use? Since nothing in Excel can be more precise than 1E-15, we will keep looping until term *k* is within this precision of term *k-1*.

(a) Open a new VBE module and enter this function:

```
Function MacExp(x)
    Const Precision = 0.000000000000001
    MacExp = 0
    Term = 1
    k = 1
    Do While Term > Precision
      MacExp = MacExp + Term
      Debug.Print k; Term; MacExp
      Term = Term * x / k
      k = k + 1
      If k > 100 Then
        MsgBox "Loop aborted, k > 100"
        Exit Do
      End If
    Loop
End Function
```

The Debug statement is to check the operation of the function; it can be deleted or "commented out" when not needed. It prints results in the Immediate Window. Note that we have an "escape hatch;" if k gets above 100, we jump out of the loop after displaying a message. It is a good idea to use this technicality to avoid unending loops. The 100 limit will be too small for values of x much over 25.

◢	A	B	C
1	Maclaurin Series for Exp(x)		
2			
3	x	=MacExp(x)	=EXP(x)
4	2	7.389056098930650	7.389056098930650
5	Difference	0.000000000000000	

Figure 9.14

(b) Set up Sheet6 as in Figure 9.14 to test the function.

(c) Save the worksheet.

Did you find any problems? How about $x = -4$? Our function gives an answer of 1 for any negative x value. Look at the loop condition. What will be the sign of *Term* after the first iteration? We can solve this by using Do While Abs(Term) > Precision.

Variables and Data Types

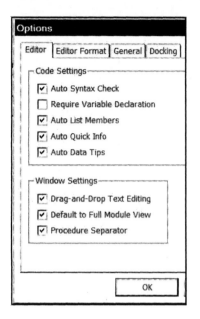

Unlike most computer languages, all dialects of BASIC allow the programmer to use variables without first declaring them. While this feature slightly speeds up the coding process, it has the major disadvantage that typo errors can go undetected. What would happen in the last exercise if you had mistakenly typed *MacExp = MacExp + Team* where the variable *Term* is misspelled as *Team*. Since *Team* is not mentioned elsewhere, its value is zero and the final result would have been zero. It is easy to make silly typos. The problem is avoided by making variable declarations mandatory. We can do this with the use of the *Option Explicit* statement at the start of the module, or by opening the VBE *Tools | Options* dialog and on the *Edit* tab checking the *Require Variable Declaration* box. You are encouraged to use the latter. With this in place, all variables must be declared before being used. We use the DIM statement for this purpose. The code for the last exercise would then begin with:

```
Function MacExp(x)
    Dim Term, k
    Const Precision = 0.000000000000001
```

MacExp is already defined by the function header, and Precision gets defined by the *Const* statement. Now if you use *Team* when *Term* was needed, VBA will issue a warning.

There is yet another difference from other programming languages. In languages such as FORTRAN and C, it is not sufficient merely to name the variable; the programmer must also state its data type. In this example we would need to define *k* as an integer variable and term as a floating-point variable. We could do this by coding *Dim Term As Double, k As Integer.* We would use *Double* rather than *Single* to get the maximum precision. When the data type of variables is not declared, VBA uses a special data type called the variant. This is acceptable for the simple functions shown in these examples, but in general one should declare data types. It should be noted that variant data types are memory hogs. You may wish to use Help to find out more about this topic, especially the permitted range of values for Integer, Short, Long, Single and Double.

Input-Output of Arrays

In the function EquivR in Exercise 8, we saw how an input array may be processed by a VBA function. That was a one-dimensional array. Figure 9.15 shows a function that processes a two-dimensional array. The objective is to find the sum of quotients a/b for each row in the table. So we have 16/8 + 25/5... giving 32. The function is called with =SUMQUOT(A4:B10). Note that it will work with any number of rows.

◢	A	B	C	D	E	F	G	H
1	A 2-Dimensional Table							
2								
3	**a**	**b**	Option Explicit					
4	16	8	Function SumQuot(myRange As Range)					
5	25	5	Dim j As Integer					
6	27	9	For j = 1 To myRange.Rows.Count					
7	16	4	SumQuot = SumQuot + myRange(j, 1) / myRange(j, 2)					
8	6	3	Next j					
9	32	4	End Function					
10	24	3						
11	SumQ	32						

Figure 9.15

The reader is encouraged to experiment with this function on Sheet7. In the next Exercise we look at a function that outputs an array: it is used to put results in a numbers of cells so it is an array function. Within the function we have an 3-by-1 array into which data is placed, and then we pass that array to the function-name in the final statement.

Exercise 10: An Array Function

For the subscripts of arrays, VBA uses a counting system starting with zero. This applies to arrays declared within the code as with *Temp* in the current UDF. However, for an Excel range passed to the UDF, the first element is referenced with a subscript value of 1. See the code in Figure 9.15.

To avoid this type of confusion, you can use *Option Base 1* at the top of the module to use a counting system that starts with 1 for all arrays.

For our example we will write a function to find the real roots of a quadratic equation $ax^2 + bx + c = 0$.

(a) In Chap9.xlsm, insert Module 7 and enter this code:

```
Function Quad(a, b, c)
  Dim Temp(3)
  d = (b * b) - (4 * a * c)

  Select Case d
    Case Is < 0
      Temp(0) = "No real"
      Temp(1) = "roots"
      Temp(2) = ""
    Case 0
      Temp(0) = "One root"
      Temp(1) = -b / (2 * a)
      Temp(2) = ""
    Case Else
      Temp(0) = "Two roots"
      Temp(1) = (-b + Sqr(d)) / (2 * a)
      Temp(2) = (-b - Sqr(d)) / (2 * a)
  End Select
  Quad = Temp
End Function
```

◢	A	B	C	D
1	Quadratic Equation Solver			
2				
3	$ax^2 + bx + c = 0$			
4		a	b	c
5	Coeff	1	-5	6
6	Roots	Two roots	3	2

Figure 9.16

(b) Using Figure 9.16 as a guide, test this in Sheet8. Remember when you have selected B6:D6 and type =Quad(A5,B5,C5) you must use Ctrl + ⇧ Shift + ↵ to commit the formula.

(c) Save the workbook.

The statement *Dim Temp(3)* establishes a 3-by-1 array. Note that VBA arrays use an indexing system that begins with zero. If this is not acceptable, you can add the code Option Base 1 before the function header.

Using Functions from Other Workbooks

The user-defined functions we have created have been used in the workbook in which they were coded. Every function in a workbook is available from any sheet in it. A number of procedures permit us to use a function from another workbook.

1. The least efficient method is to copy the function to the new workbook using either Copy&Paste or, within the VBE File menu, Export and Import.

2. The user-defined functions of an open workbook are available in other workbooks. Thus if Chap9.xlsm is open, then in a second workbook we may code, for example, =Chap9.xlsm!Quad(...). We could either type this formula or use the Insert Function tool to locate the function in the User-defined category. Frequently used macros may be conveniently kept in a file called Personal.xlsb stored in the Xlstart folder. Normally, one hides this workbook in Excel (with *View | Window | Hide*) before it is saved. Files in Xlstart automatically open when Excel starts.

3. Finally, we look at the steps needed to make an add-in from Chap9.xlsm once all the functions have been coded and thoroughly tested.

The file extension *xlsb* is used for binary files, which are smaller and faster to open than *xlsx* files.

(i) Rename and lock the project. Open the project properties dialog either from the shortcut menu found when you right click a member of Chap9.xlsm in the Project Window, or by using the Tools menu command. By default, all projects are named VBAProject; it is better to give it a unique name such as *Chap9.* If you are distributing the add-in, you may wish to lock the code and add a password so others cannot see or alter it.

(ii) Add file properties: Return to Excel, use *Office | Prepare | Properties.* In the dialog, enter descriptive information about your add-in.

(iii) Save in Add-in format: Use *Office | Save As | Other* to open the Save As dialog. In the Type box near the end of the list, select Excel Add-in (.xlam). You could also save an Excel 97-2003 Add-in. The file will be saved with the extension xla.

(iv) Install the Add-in: You, and others to whom you give the file, can install the add-in using the Add-in Manager found in *Office | Excel Options.* The Manager box should read *Excel Add-Ins*; click the Go button. Use the Browse command to locate and install the add-in. The add-in will display the descriptive information you added in step (ii). Now you will be able to use all the functions in the add-in as if they were in the current file.

A number of companies market Microsoft Excel add-ins. For example, one add-in contains functions for performing mass-mole chemistry calculations. You may find some shareware add-ins by searching the World Wide Web.

Problems

1. Alter the Quad function in Exercise 10 in such a way that it may be called with =Quad(A1:C1).

2. *The pressure drop in a pipe of length L ft and diameter D ft with an average water velocity of V ft/sec is given by the Darcy-Weisbach equation

$$\Delta P = f\,\frac{L}{D}\,\frac{V^2}{2g}$$

where f is the friction factor (generally taken as 0.02) and g is the acceleration due to gravity (32.2 ft/sec^2). Write a UDF with the header Function(Length, Diameter, Flow, Friction) where *Flow* is the average volume flow in ft^3/sec. Test it against results obtained with worksheet formulas and/or an Internet site.

3. *Write a UDF to calculate the resultant of two vectors; see Figure 9.17. Cells B6:C6 call the function with =ForceVector(B4:C4,B5:C5) as an array function. The formulas to solve this are as shown in the figure.

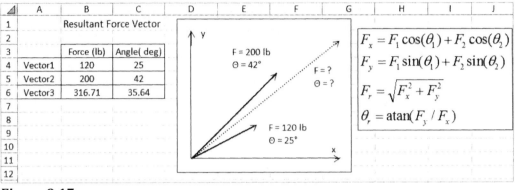

Figure 9.17

4. *Write an array function with the header Function SciNum(Number) that accepts a number and returns the significant and exponent in two separate cells as shown in Figure 9.18.

	A	B	C
1	Scientific Numbers		
2			
3	Number	SciNum	Power
4	45.678	4.5678	1
5	123.99	1.2399	2
6	293456	2.93456	5
7	4.342	4.342	0
8	1010.29	1.01029	3

Figure 9.18

5. Refer to Problem 4 in Chapter 4. Write a UDF with the header Function Trough(length, radius, height) that returns the volume of water in the trough. The arguments *length* and *radius* are in feet, while *height* is in inches. The result should be rounded to the nearest gallon. Write another UDF where *height* refers to the water depth—the length of the wet part of the dip stick rather than the dry part.

6. *A range (vertical or horizontal) in a worksheet contains the magnitude of some vectors. Write a UDF that accepts the range and returns the magnitude of the resultant vector using $F = \sqrt{\sum f_i^2}$.

7. Rework Problem 9 of Chapter 2. Write a UDF with the header Function SolExt(mass, Kd, Vw, Vs, n) to return the mass of solute in the water after *n* extractions. The arguments are: *mass* is the amount of solute in the water phase before extraction, *Kd* is the distribution constant, *Vw* is the volume of water, and *Vs* is the volume of solvent.

8. Developing series expansions for π seems to have been a favorite pastime for mathematicians of old. Two of historic interest are as follows.

$$\text{Gregory-Leibniz:} \quad \frac{\pi}{4} = 1 - \frac{1}{3} + \frac{1}{5} - \frac{1}{7} + \frac{1}{9} = \sum_{n=0}^{\infty} \frac{(-1)^n}{2n+1}$$

$$\text{Wallis:} \quad \frac{\pi}{2} = \frac{2 \cdot 2}{1 \cdot 3} \cdot \frac{4 \cdot 4}{3 \cdot 5} \cdot \frac{6 \cdot 6}{5 \cdot 7} = \sum_{n=1}^{\infty} \frac{(2n)^2}{(2n-1)(2n+1)}$$

Write two UDFs and compare the results of the Gregory-Leibniz series with that of Wallis for *n* = 1000 and *n* =1 × 10^6. Do not try *n* greater than one million; to suggest 10 years ago that one lets even these small pieces of code loop a million times would have raised eyebrows! Neither series is of practical use, but they make excellent programming challenges. These functions may slow down your workbook if you leave *n* with a large value.

9. The Antoine equation is as follows.

$$\log_{10}(p^*) = A - \frac{B}{T+C}$$

where p^* is in mmHg and *T* is the temperature in degrees Celsius. Write a UDF with the header Antoine(A, B, C, P), which will return the boiling point (rounded to one decimal) of the liquid when the external pressure is *P*. Do not use any worksheet functions in your code. Data to test your formula: for benzene *A*, *B*, and *C*, are 6.90565, 1211.033, and 220.790 respectively, and its normal boiling point is 80.1°C.

10. You are planning to write code to perform matrix algebra on the numbers in a range. Your code will require that the range be square, that is, that the number of rows must equal the number of columns. In preparation for this write a UDF with header Function IsSquare(myRange) that will return TRUE or FALSE.

11. Write a UDF with the header Function SumOfDiagonal (myRange), which returns the sum of the diagonal elements in

myRange if it is square, or a suitable error message otherwise. This UDF should call the one above in a statement equivalent to Test = IsSquare(myRange). The sum of the diagonal elements is often called the *trace* of the matric.

12. Write an array UDF with the header Function NormalArray (myRange) which will return a new array equal to the input array normalized by dividing each element by the sum of the squares of all elements. In the code, declare a 100 by 100 array to hold the calculated values. Do not bother to redimension this. It will be simpler if you use Option Base 1. For your first attempt you may use a worksheet function but then challenge yourself to make it work without the Excel function.

13. Write a UDF that takes in a two-column array (up to 100 by 2) and returns an array such that, in each row, the left column has the larger value and the right column has the smaller.

14. A ball is thrown at 50 mph at an angle of 30° to the horizontal. How high will it be when it has traveled 50 feet horizontally? Write a UDF to solve this using the equation below. Also use it to make a plot of the ball's path. Do your results show that the maximum distance is achieved with a 45° angle?

$$y = \frac{\sin\theta}{\cos\theta}x - \frac{g}{2v^2\cos^2\theta}x^2$$

[1] M. R. Cullen, *Mathematics for the Biosciences*, PWS Publishers, Boston, 1983, (page 344).

15. A cylinder of radius r has a cone attached to its base. The sides of the cone make an angle θ to the axis of the cylinder. The object has a volume V. The surface area[1] of the object is given by

$$S = \frac{2V}{r} + \pi r^2 \left(\csc\theta - \frac{2}{3}\cot\theta \right)$$

Write a UDF with the header Function SurfaceArea(r, V, theta) to compute the value of S. You might show by calculus that S is minimized when *theta* = cos^{-1}(2/3) or about 48.2°. Use the function to plot S vs. theta and confirm this value. Think of a way in which you could perform a check on your function.

VBA Subroutines

Early in the history of personal computers, software developers added scripting languages to their applications. This allowed users to write *macros.* The term is short for macro-instruction; so while the command *Run MyMacro* is considered a single instruction it probably invokes a large number of instructions. This makes macros extremely useful for repetitive operations, but in the case of VBA in Excel we can extend the use to other purposes.

Macros can be either recorded or coded. A recorded macro can be fine tuned by editing. Alternatively, one can record a short macro to learn the syntax of a particular operation and paste the result into a larger subroutine.

The macros here have been kept fairly simple so the reader is advised to have only the Chap10.xlsm workbook open when running them. Otherwise you may have unexplained results.

The companion website has supplementary material about VBA subroutines.

The words *macro* and *subroutine* are almost synonymous. If you record a macro called *Alpha*, VBA will create a subroutine with the same name. If you use the command *View Macros* you will see *Alpha* within the macro list. Perhaps the subroutine *Alpha* calls subroutines *Beta* and *Gamma*, and function *Kappa*. Then the entire group constitutes the macro *Alpha*. Generally, these would be kept on a single module sheet.

There is a special class of macros called *event macros* that are associated with either a specific sheet or with the entire workbook. They are automatically executed whenever a specific event occurs. For example, there could be a before-close macro, which saves the workbook whenever the user attempts to close it. We shall not be investigating these further.

Exercise 1: Recording a Macro

For this exercise our scenario is: Each day you receive a file similar to Figure 10.1. You wish to copy and paste this to another workbook and perform some operations on it. For the exercise we shall add a sorting macro, which can be called with Ctrl+⇧Shift+Q. This is a very simple operation, but it will demonstrate the procedures needed to record a macro.

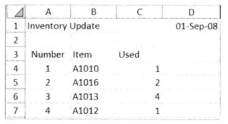

◢	A	B	C	D
1	Inventory Update			01-Sep-08
2				
3	Number	Item	Used	
4	1	A1010	1	
5	2	A1016	2	
6	3	A1013	4	
7	4	A1012	1	

Figure 10.1

(a) Download the file Chap10Data.xlsx from the companion website. Copy and paste everything to Sheet1 of a new workbook and save it as a macro-enabled workbook named Chap10.xlsm.

(b) Begin the macro recorder using *Developer | Code | Macro Recorder*. Complete the record macro dialog as shown in Figure 10.2 and click OK. The shortcut key label will change from *Ctrl+* to *Ctrl+Shift+* as you type in the box.

Record Macro ? ✕

Macro name:
> SortData

Shortcut key:
> Ctrl+Shift+ Q

Store macro in:
> This Workbook ▾

Description:
> Sort incoming inventory usage

OK   Cancel

Figure 10.2

Observe in the *Code* group that the *Record Macro* command has changed to *Stop Recording* with a blue rectangular icon and a similar icon is displayed next to *Ready* in the status bar. Only actions are recorded, so there is no merit in rushing the process.

(c) Click anywhere within the data (for example cell A6) and use the command *Data | Sort & Filtering | Sort.* Fill in the sort dialogue as shown in Figure 10.3 and click OK. We sort first by *Item,* then by *Used.*

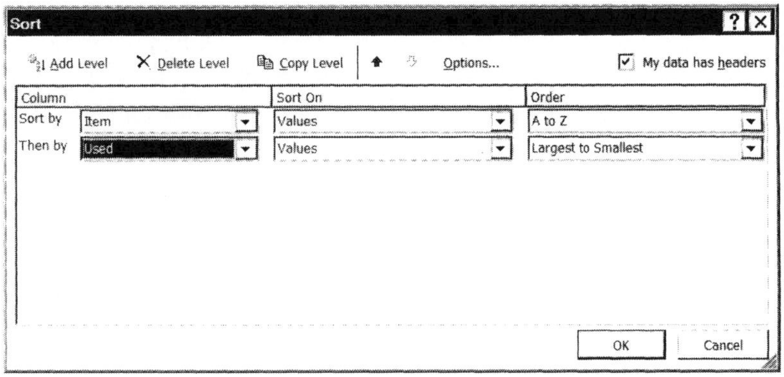

Figure 10.3

(d) If you look at the data, you will see that it is now sorted. We can turn the recording off either by opening the *Developer* tab or by clicking the blue icon in the status bar.

(e) Use *Developer | Code | Visual Basic* and open the module for your file. We will not investigate every line in this code, but the meaning of some of it should be fairly obvious.

Now we will show how the macro can be used in production.

(f) Delete all the data on Sheet1 and re-copy the original data from Chap10Data.xlsx.

(g) Using either $\boxed{\text{Ctrl}}$+$\boxed{\text{⇧ Shift}}$+Q or *View | Macros | View Macros* run the macro *DataSort*. The data gets sorted.

(h) Save the workbook as Chap10.xlsm using the command *Office | Save As | Excel Macro-Enabled.*

Computing Subroutines

Before coding a calculation macro you should spend time planning its algorithm—a specific set of instructions for solving a problem. Some programmers like to do this with flow charts, while others prefer writing pseudocode (informal VBA without too much regard to syntax). For further information on these topics consult an introductory program book or consult one of the websites listed on the companion website.

The process of coding a VBA macro to carry out a computation is the same as programming in any other language with the important exception that generally the input and output is from, and to, worksheet cells. This saves a great deal of coding.

Using a macro for the calculations gives us extra work but adds flexibility. Suppose we wish to perform an iterative process varying *h* from 1 to 100 in steps of 0.1. Doing this on a worksheet would take 1000 cells. With a macro we could elect to have output when *h* was an integer or a multiple of 10; or only have the final result sent to the worksheet.

It is strongly recommended that, when developing your own macro, you do not try to code it all in one go. Write part of the code and have the function return an intermediate value. Use debugging statements (shown later) if needed. Add a bit more code and so on to completion.

Notes on the VB Editor

1. If you run code with an error that causes the program to "hang," the offending line will be highlighted in the module. You will not be able to do anything on the worksheet while this condition lasts. The module can be reset within VBE with the command *Run/ Reset* or with the reset tool; see Figure 10.4.

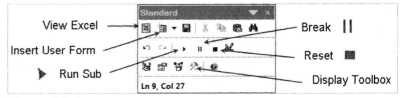

Figure 10.4

2. If your subroutine goes into an infinite loop, use [Ctrl] + [Break] or [Esc] to terminate it. You can do this from Excel or from VBE. There also is a tool on the VBE toolbar to break a macro.

3. From the Excel window, a macro may be run using either *Developer | Code | Macros* or *View | Macros | View Macros* to open a dialog box that lists all the available macros. You may also start macros from within VBE by: (i) Using the *Run Sub* command in the *Run* menu; (ii) pressing [F5]; or (iii) using one of the tools in Figure 10.4. You can also tap [F8] and run a macro one line at a time. Always ensure the current worksheet is active before using these tools.

Exercise 2: A Computing Macro

$$F_x = F_1 \cos(\theta_1) + F_2 \cos(\theta_2)$$

$$F_y = F_1 \sin(\theta_1) + F_2 \sin(\theta_2)$$

$$F_r = \sqrt{F_x^2 + F_y^2}$$

$$\theta_r = \text{atan}(F_y / F_x)$$

For our demonstration we shall re-do the calculation of Problem 3 in Chapter 9—finding the resultant of two force vectors. There is no real advantage in using a macro here, but we want to start with an example that is not too complex. The simple algorithm is shown in the sidebar.

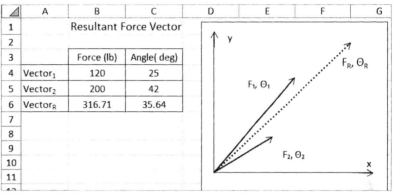

⊿	A	B	C	D	E	F	G
1		Resultant Force Vector					
2							
3		Force (lb)	Angle(deg)				
4	Vector$_1$	120	25				
5	Vector$_2$	200	42				
6	Vector$_R$	316.71	35.64				
7							
8							
9							
10							
11							

Figure 10.5

There is a simple algorithm for this macro:
Get input data for worksheet.
Compute Fx and Fy.
Compute Fr and θr.
Put values on worksheet.

In a hurry? As an alternative to typing, download the file *Chap10Vector.bas* from the companion website and in the VBE use the command *File / Import* to load the macro.

(a) Prepare a worksheet on Sheet2 of Chap10.xlsm as in Figure 10.5, leaving B6 and C6 blank. The chart is optional.

(b) Open the VBE using *Developer | Code | Visual Basic* and insert a new module (Module2) on the Chap10.xlsm project. Type the code shown in Figure 10.6, omitting the line numbers that are used in the discussion below.

(c) Return to Excel. Use *View | Macros | Macros | ViewMacros* and run the macro *ResultantForceVector*. If you do not get the correct results, lines 17 and 22 give examples of how to get debugging information; just remove the leading apostrophe so they are no longer comments. Do a Save.

Comments on the code:
(i) Lines 1–5: These are discussed just below the exercise.
(ii) Lines 6–9: We have made no provisions to ensure the macro is run only from Sheet2. This code gives the user a chance to back out of the macro.
(iii) Line 11: Defining π this way is superior to entering a value.
(iv) Lines 13–17: This is where the program gets its input values from the worksheet. Note the use of meaningful names. Line 17 (with the apostrophe removed) is an example of debugging code.
(v) Lines 19–22: Here we do the intermediate calculations. Line 22 is an interim output and can be used for debugging.
(vi) Lines 24 and 28: We compute the final results.

(vii) Lines 30–32: These are the output statements that put data on the worksheet. Line 28 could have begun with *Range("B6")*... but the Cells (note the plural) method is a useful one to know as we see in the next exercise.

```
1    Option Explicit
2    'To compute the resultant of two vectors
3    Public Sub ResultantForceVector()
4      Dim Answer, Pi
5      Dim Force1, Force2, Theta1, Theta2, ForceX, ForceY, ForceR, ThetaR
6      Range("B6").Select
7      Const MyQuestion = "Is B6 where the resultant forces goes?"
8      Answer = MsgBox(MyQuestion, vbYesNo + vbQuestion, "Vectors")
9      If Answer <> 6 Then Exit Sub
10
11     Pi = 4 * Atn(1)          'VBA calculation for Pi, 1 is in radians
12
13     Force1 = Range("B4").Value
14     Theta1 = Range("C4").Value * Pi / 180 'convert degrees to radians
15     Force2 = Range("B5").Value
16     Theta2 = Range("C5").Value * Pi / 180
17     'Debug.Print Force1; Theta1; Force2; Theta2
18
19     'calculate ForceX and ForceY
20     ForceX = Force1 * Cos(Theta1) + Force2 * Cos(Theta2)
21     ForceY = Force1 * Sin(Theta1) + Force2 * Sin(Theta2)
22     'MsgBox "ForceX " & ForceX & " ForceY" & ForceY
23
24     'calculate R
25     ForceR = Sqr(ForceX ^ 2 + ForceY ^ 2)
26
27     'calculate Theta
28     ThetaR = Atn(ForceY / ForceX) * 180 / Pi 'convert radians to degrees
29
30     'place results in cells
31     Cells(6, 2).Value = Round(ForceR, 2)
32     Cells(6, 3).Value = Round(ThetaR, 2)
33   End Sub
```

Figure 10.6

All the *.Value* phrases are optional, so code such as *Force1 = Range("B4")* would work. However, most VBA experts recommend the use of *Value*.

If the values in B4:C5 are changed, the resultant information is not updated until the macro is rerun. This will be a good example of where a "change event" macro could be used to display a message whenever the input data was altered.

Public or Private?

The header of our macro begins with the word *Public*. This means the macro can be seen from any sheet in any open workbook. The word is optional since it is the default setting. If *Public* is replaced by *Private*, the macro is no longer available through the *View* tab, but it may be run from a control placed on the worksheet. In Exercise 5 we see how this may be accomplished.

Name That Variable

We have used names like *Force1*, *ForceX* rather than *F1* and *FX*. The additional typing effort will be well invested. The meaningful names let you think about the algorithm. If you or someone else revisits the code months from now, the long names will aid in understanding what is going on. You may not find this argument convincing with such a simple program, but with longer programs it becomes more important.

Remember from the previous chapter that some words have special meanings in VBA. These *keywords* cannot be used for naming variables. So although *Count* is not actually a keyword, the cautious programmer might use *Kount*.

Simple variables like *j*, *k*, *n* for counters are fine, as are *x* and *y* when working with Cartesian coordinates.

We have used *Option Explicit* and hence need the Dim statements. This was discussed in the previous chapter. As the code gets longer, typo errors become more difficult to spot, so while this feature was not essential with UDFs, with subroutines it is indispensable.

Exercise 3: Bolt Hole Positions

A milling machine is set up to drill holes in a circular pattern. The variable *Bcount* gives the number of holes while *Diam* is the diameter of the drilled circle. The first hole is drilled at an angle *Alpha* from a reference mark on the circle. Our task is to compute for each hole, the *x, y* coordinate values where 0, 0 are the coordinates of the circle's center. These values will be fed into the milling machine. The relevant equations are

$$\theta = \frac{n \cdot 360}{Bcount} + Alpha$$

$$x = \tfrac{1}{2} \, Diam \cos(\theta); \quad y = \tfrac{1}{2} \, Diam \sin(\theta)$$

◢	A	B	C	D
1	Bolt circle hole calculation			
2				
3	Bolt hole diameter		6	inch
4	Offset angle		22	degree
5	Number of bolt holes		7	
6				
7	Bolt hole #	x [in.]	y [in.]	
8	1	2.78	1.12	
9	2	0.86	2.88	
10	3	-1.71	2.46	
11	4	-2.99	0.19	
12	5	-2.02	-2.22	
13	6	0.48	-2.96	
14	7	2.61	-1.47	
15				

Figure 10.7

(a) On Sheet 3 of Chap10.xlsm, copy the entries from Figure 10.7 but leave rows 8 to 15 blank.

(b) The flow chart for our macro is shown in Figure 10.8. Open VBE, add a third module, and enter the subroutine shown in Figure 10.9. It is also available on the website as Chap10Bolt.bas.

(c) Run the macro a few times with various settings of the three input parameters.

(d) Save the workbook.

In line 8 we take care to open the correct worksheet. Then we delete any existing output. Recording a macro helped the author recall the syntax for lines 10 and 11. Input values are taken from the worksheet in lines 14–17. Then we have a FOR loop that iterates *bcount* times calculating and outputting values of x and y to the worksheet. Note the use of Cells to place data on varying rows. The syntax is Cells(row, column).

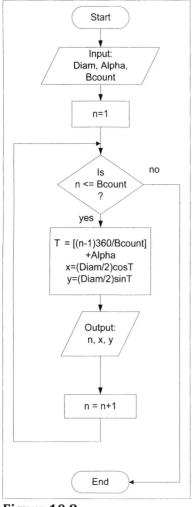

Figure 10.8

```
1    Option Explicit
2    Public Sub BoltHoleCircle()
3    Dim Pi, Diam, Alpha, Bcount, Theta, x, y, n
4    'declare constants
5        Pi = 4 * Atn(1)
6
7    ' open Sheet3 and clear old data
8        Sheets("Sheet3").Select
9        Range("A8:C8").Select
10       Range(Selection, Selection.End(xlDown)).Select
11       Selection.ClearContents
12       Range("A8").Select
13
14   'input dia, alpha, bcount
15       Diam = Range("C3").Value
16       Alpha = Range("C4").Value
17       Bcount = Range("C5").Value
18
19   'Do calculations
20       For n = 1 To Bcount
21           Theta = ((n - 1) * 360 / Bcount) + Alpha
22           Theta = Theta * Pi / 180 'convert to radians
23           x = Round(Diam / 2 * Cos(Theta), 2)
24           y = Round(Diam / 2 * Sin(Theta), 2)
25       'output n, x, y
26           Cells(7 + n, 1).Value = n
27           Cells(7 + n, 2).Value = x
28           Cells(7 + n, 3).Value = y
29       Next n
30   End Sub
```

Figure 10.9

Exercise 4: Finding Roots by Bisection

We next demonstrate a VBA implementation of the bisection method. The main subroutine *Bisection* calls a UDF named *MyFunction* to evaluate the function to be solved at specified *x* values. This will give us the flexibility of being able to re-code only *MyFunction* when we wish to solve another equation.

A brief explanation of how the bisection method works. The equation to be solved is f(x) = 0. By trial and error we have found that values of f(a) and f(b) have opposite signs. Clearly, the solution to f(x) = 0 lies in the interval $a < 0 < b$. We now find the midpoint ("bisect") of this interval $m = (a + b)/2$ and evaluate f(m). If f(m) has the same sign as f(a), then the root lies in $m < x < b$ as illustrated on the left of Figure 10.10. Alternatively, if f(m) and f(b) have the same sign, the root lies in $m < x < b$ (see the right side of the figure). We can now bisect either m and b or m and a to get a smaller interval. We may repeat this until we are close enough to zero to satisfy our precision needs.

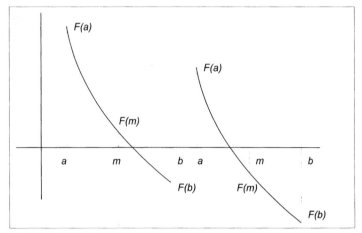

Figure 10.10

This allows us to develop an algorithm for finding a root of f(x):
    Start with values of a and b such that f(a) and f(b) have opposite signs
    Loop until the required accuracy is achieved
        Find the midpoint $m = (a + b)/2$
        If f(m) and f(b) have opposite signs
            give a the value of m
        Else
            give b the value of m
        End if
    End loop.

We will solve the transcendental equation $\exp(x) - \sin(x) = 0$. The graph of this (Figure 10.11) shows there is one root between 0 and 1, and another between 2.5 and 3.5. These values were arbitrarily read from the chart.

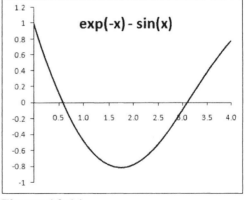

Figure 10.11

◢	A	B	C	D	E	F
1	Bisection Method					
2						
3	a	b	midpoint	f(a)	f(b)	f(midpoint)
4						
5						

Figure 10.12

The code is available as Chap10Bisection.bas on the companion website.

(a) On Sheet 4 of Chap10.xlsm, set up a worksheet similar to that in Figure 10.12. You may wish to experiment using worksheet functions to implement the Bisection algorithm before moving on to the VBA coding. The formula =SIGN(A1) returns the value –1, 0 or +1 depending on the value in A1 (negative, zero, or positive).

(b) On the Chap10.xlsm project in the VBE, add Module 4. Enter the code shown in Figure 10.13. A statement ending with a space followed by an underscore is continued on the next line. It may be entered as shown or as a single line without the underscore.

(c) Experiment with various input values. Then modify the function *Kappa* to solve $f(x) = 4^x + 5^x - 100$, which has a root between 0 and 3. Make a plot to confirm this.

(d) Save the workbook.

```
Option Explicit

Function Kappa(x) As Single
  Kappa = Exp(-x) - Sin(x)
End Function

Public Sub BisectionMethod()
   'declare variable types
   Dim a As Single, b As Single,
   Dim midpoint As Single
   Dim f_a As Single, f_b As Single,
   Dim f_m As Single
   Dim Kount As Integer, MyRow As Integer

 ' Select worksheet and clear old data
   Sheets("Sheet4").Select
   Range("A4:F4").Select
   Range(Selection, _
           Selection.End(xlDown)).Select
   Selection.ClearContents

 ' get starting values a and b
   a = InputBox("For f(a) give value of {a}" _
   ,"Bisection Method")
   b = InputBox("For f(b) give value of {b}" _
   ,"Bisection Method")
   f_m = Kappa((a + b) / 2)
 ' test that f(a) and f(b) have opposite signs
   If Sgn(Kappa(a)) = Sgn(Kappa(b)) Then
     MsgBox "f(a) and f(b) _
       have the same sign"
     Exit Sub
   End If
 'initialize Kount

    Kount = 1
    'the algorithm, precision set to 0 ± 1.0E-6
    Do While Abs(f_m) > 0.000001
      midpoint = (a + b) / 2
      f_a = Kappa(a)
      f_b = Kappa(b)
      f_m = Kappa(midpoint)

    'output a, b, midpoint, f_a, f_b, f_m
      Cells(Kount + 3, 1) = a
      Cells(Kount + 3, 2) = b
      Cells(Kount + 3, 3) = midpoint
      Cells(Kount + 3, 4) = f_a
      Cells(Kount + 3, 5) = f_b
      Cells(Kount + 3, 6) = f_m

    'Compare sign of f(a) with sign of f(b)
      If Sgn(f_m) <> Sgn(f_b) Then
        a = midpoint
      Else
        b = midpoint
      End If

      Kount = Kount + 1
    'Check the number of iterations
      If Kount >= 50 Then
          MsgBox "Not converging"
          Exit Sub
      End If
    Loop
    MsgBox "Solution at x = " & _
        midpoint & " when f(x) = " & f_m
End Sub
```

Figure 10.13

When a single statement of code needs to be continued on a second line, we make the break with a space followed by an underscore. You may enter the code into the VBE window in this way, or you may join the lines together omitting the underscore.

Exercise 5: Using Arrays

Variable arrays can be used to organize the way input/output data is stored. The worksheet in Figure 10.16 represents a table of soil contamination levels of carbon tetrachloride. In this exercise the variable array for each soil contamination level is named *SCL*. We declare this as SCL(20) using the Dim command; see Figure 10.14. The number 20 indicates the size of the array. We can refer to SCL(1) through SCL(20). An integer variable *i* will be used in a Do Loop to input all the SCL values until it lands on a blank cell. We have chosen to have the variable array *Exceed* store the index (position) of SCL values greater than 0.01 rather than the actual values. This is a useful technique with large arrays since the integer value of the index takes little storage room.

```
Public Sub SoilContamination()
'declare array variables
Dim SCL(20)
Dim Exceed(10)

'initialize counting variables
  row_count = 1
  i = 1
  h = 0

'start Do Loop; test for a value in the cell
  Sheets("Sheet5").Select
  Range("B3").Select
  Do While ActiveCell.Value > 0
  'input Soil Contamination Level
    SCL(i) = ActiveCell.Value
    If SCL(i) > 0.01 Then    ' test for level
      h = h + 1
      Exceed(h) = i
      ActiveCell.Font.Bold = True
    End If
    'move down 1 row
    ActiveCell.Offset(1, 0).Activate
    i = i + 1
  Loop

'output h (number of occurrences _
      exceeding 0.01)
  Range("D3").Select
  ActiveCell.Offset(-1, 1).Value = _
  "There are " & h & " samples _
          with CTC > 0.01 ppm"
For j = 1 To h
  'output SCL's that exceed 0.01
  high = Exceed(j)
  ActiveCell.Value = SCL(high)
  ActiveCell.Offset(0, 1) = "ppm"
  'move down 1 row
  ActiveCell.Offset(1, 0).Activate
Next j

Const MyQuest = _
      "Ready to reset the worksheet?"
Const MyText = "Soil Contamination"
Answer = MsgBox(MyQuest, vbYesNo _
    + vbQuestion, MyText)
If Answer = 6 Then Clearworksheet
End Sub
```

Figure 10.14

So that the reader can experiment with different data sets, *SoilContamination* ends with a call to *Clearworksheet.* Its code is shown in Figure 10.15.

```
Sub Clearworksheet()
   Range("B3").Select
   Do While ActiveCell.Value > 0
      ActiveCell.Font.Bold = False
      ActiveCell.Offset(1, 0).Activate
   Loop
   Range("E2").Clear
   Range("D3").Select
   Do While ActiveCell.Value > 0
      ActiveCell.Clear
      ActiveCell.Offset(0, 1).Clear
      ActiveCell.Offset(1, 0).Activate
   Loop
End Sub
```

Figure 10.15

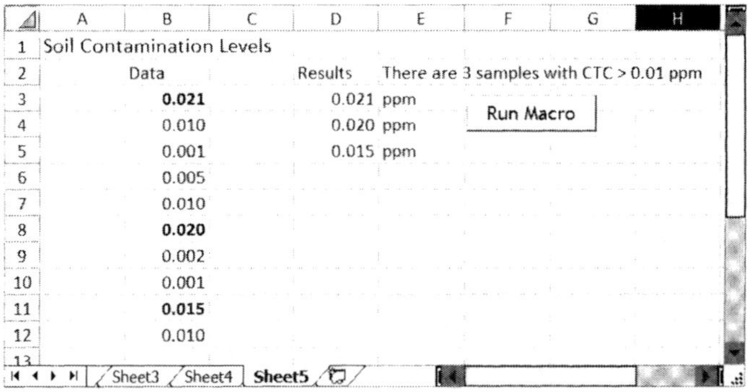

Figure 10.16

(a) Open Chap10.xlsm and on Sheet5 construct the worksheet shown in Figure 10.16. Ignore the bold format and everything after column C except the text in D2. The other entries come from the macro, and the box will be discussed in the next exercise.

(b) On Module 5 in VBE, code the macro consisting of the two subroutines in Figures 10.14 and 10.15.

(c) Run the macro and experiment with different data.

Suppose you have an array defined by **MyArray(4,5)** and the array elements get values through some calculation. Now you want to display these on the worksheet. This code will serve that purpose:

```
Range("A10").Select
ActiveCell.Resize(4,5).Value=MyArray
```

Exercise 6: Adding a Control

We have been running macros using from Excel with either *Developer | Code | Macros* or *View | Macros | View Macros*, or from VBE with either the Run command or using a tool.

Another way is to add a control to the worksheet. This is very convenient and ensures you are running the macro that matches the active sheet. It is also an excellent way to have another user, perhaps one less familiar with Excel, run a macro in a workbook you have developed.

(a) Open Sheet5 of Chap10.xlsm. Use *Developer | Insert* and click the first tool in the *Forms* group; see Figure 10.17. The mouse pointer turns to a cross (+); drag the mouse to outline a small rectangle about one column wide by two rows deep.

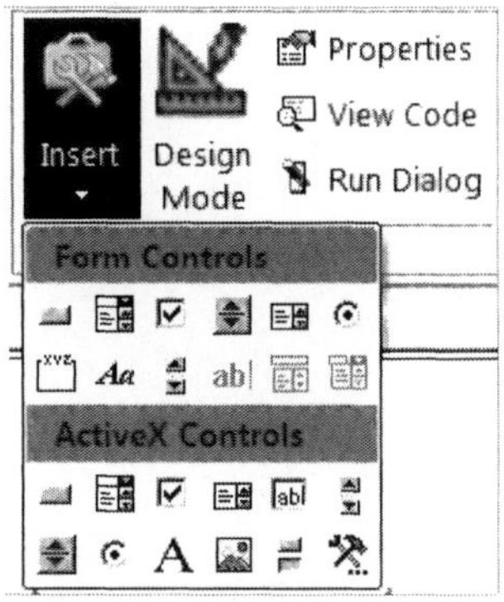

Figure 10.17

(b) When you release the mouse button, the Assign Macro dialog opens, enabling you to select which macro the control will run; SoilContamination is the appropriate one here.

(c) Right click the button and you are presented with various options. Use Edit Text to customize the control.

(d) Now when you click the control, the associated macro will run. Save the workbook.

Exercise 7: User Forms

User forms provide a data input or output interface They can also be used to launch programs that input and output on the worksheet. The user form in Figure 10.18 is associated with a macro that calculates the height of a building. A surveyor takes three measurements: distance (d) from building, angle of elevation (θ) of theodolite and height of theodolite (h_1) above ground level; see Figure 10.19. The height of the building is given by:
$h = d \tan(\theta) + h_1.$

Figure 10.18

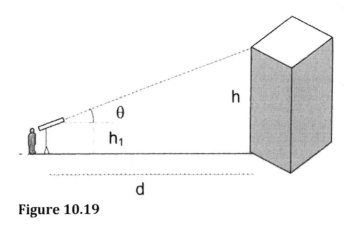

Figure 10.19

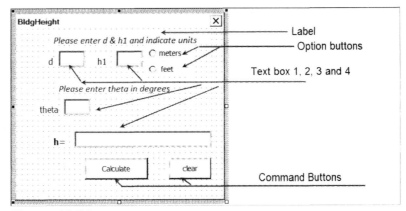

Figure 10.20

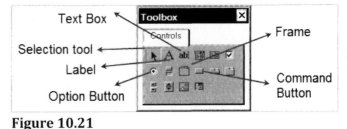

Figure 10.21

(a) We begin the exercise by designing the User Form. Learning to make a User Form is like learning to drive a car; someone must get you started telling you what all the knobs are about, but you have to practice and learn how to use them smoothly. What follows tells you the very basic information; refer to Figures 10.20 and 10.21 as you read.

(i) Begin in VBE by using the command *Insert | UserForm*. Now you have a blank form onto which we will add *controls* such as *Text labels, Text boxes, Command Buttons,* and *Option Buttons.*

(ii) We need the *Toolbox* from which to select the controls: have this displayed by using *View | Toolbox* or by clicking the crossed hammer-&-wrench icon next to the Help icon. If the mouse is allowed to hover over a tool, a screen tip displays its purpose. Do not worry that the controls do not look quite right as you add them; we fix that in step (iv).

(iii) Begin by dragging the Label tool (A) onto the form and typing the text Please enter d & h1 and indicate units. The Selection tool (arrow) lets you resize and position controls. Add the second label.

(iv) With a group of options buttons we want to be able to select and have the others automatically be deselected. To do this we need first to add a *Frame* to the form and then place the *Option Buttons* inside it.

(v) Finally, add the *Text Boxes* using the tool labeled *ab*.

(vi) Now you may want to change the appearance of some controls. Click each one (just once!) in turn and look in the Properties box; see Figure 10.22. We use this to alter such things as labels (e.g., take the frame label off), size, color, font, and so on. Make sure you click the User Form area, and in Properties change its name to BldgHeight.

Figure 10.22

(b) Now we need to write some code for the command buttons telling VBA what to do when either one is pressed. Double click each command button in turn and add the subroutines shown in Figure 10.23. These go together on one module. Command Button 1 is the Calculate button.

(c) In VBE, add Module 6 and this simple subroutine:

```
Sub get_UserFormBldgHeight()
    BldgHeight.Show
End Sub
```

(d) Finally, following the instruction in Exercise 5, add a command button to Sheet 6 of Chap10.xlm and associate it with the macro get_UserFormBldgHeight. Now you are ready to experiment with the project we have just completed.

```
Option Explicit                              'calculate building height
Public Sub CommandButton1_Click()            h = d * Tan(theta * Pi / 180) + h1

'declare variables                           'round off h
  Dim d As Single, theta  As Single          h_sigfig = Round(h, 0)
  Dim h1 As Single, h As Single             'output  h
  Dim Pi As Single, h_sigfig As Single        If OptionButton1 = True Or _
  Dim units As String                            OptionButton2 = True Then
' declare constant                               TextBox4.Value = "The building is " & _
  Pi = 4 * Atn(1)                                             h_sigfig & units & " tall"
'test for units                              Else
  If OptionButton1 = True Then                  'indicate error
    units = " meters"                           TextBox4.Value = units
  Else                                        End If
    If OptionButton2 = True Then
      units = " feet"                         End Sub
    Else
      units = "please enter d, h1 units"      Private Sub CommandButton2_Click()
    End If                                    'Clear the contents of the text boxes
  End If                                        TextBox1.Text = " "
                                                TextBox2.Text = " "
                                                TextBox3.Text = " "
'input d, theta, & h1                          TextBox4.Text = " "
  d = Val(TextBox1.Value)                       OptionButton1 = False
  theta = Val(TextBox2.Value)                   OptionButton2 = False
  h1 = Val(TextBox3.Value)                    End Sub
```

Figure 10.23

Problems

1. Use the program in Exercise 2 and make the code work so that the resultant angle will display properly in each of the four quadrants (360°). Extend the program to find the resultant of three vectors.

2. *Write a computer program to calculate the force of drag of fluid flowing over a cylindrically shaped object using the following formula

$$Fd = \frac{\frac{1}{2} \cdot Cd \cdot \rho \cdot v^2 \cdot A}{gc}$$

where Fd is the force of drag, Cd is the drag coefficient, ρ is the density of the fluid, v is the velocity, A is the frontal area of the object, and gc is the gravity constant. The program should work for air flowing over an object shaped as a convex $Cd = 1.2$ or concave $Cd = 2.3$.

The program should be a VBA subroutine with input and output in the Excel worksheet. The program should work in U.S. units. Use as input; velocity in mph, shape of the end (concave or convex), length in ft, and radius in inches, with the output in lbf. Incorporate a *Select Case* statement to handle the shape (concave or convex).

3. *Use the program in Exercise 5 so that it works for four columns of input in the worksheet.

4. Write a computer program that grades "pass" or" fail" for the following percent grades:

Student 1	Student 2	Student 3	Student 4
45	80	60	100
75	58	37	85
90	59.5		50
80			

60% and above is passing. Indicate the pass/fail in the same cell as the percent number. This Program should work with any reasonable number of columns or rows.

5. Use an array to store the value of grades in Problem 4 and output all grades greater than 85.

6. Following on from above, write a subroutine to output information on high performances (grades >=85).

Student	Exam	Grade
1	3	90
4	2	100
4	3	85

7. Review Problem 2 in Chapter 2. Write or record a macro that will clear all the input cells on the worksheet when the user clicks a control button.

8. Review Problem 8 of Chapter 5. Construct a worksheet named **Recycle** with rows 1 to 8 as in Figure 10.24. Add a button. Using the code shown below as a start, write a subroutine to generate the data in row 9 and down.

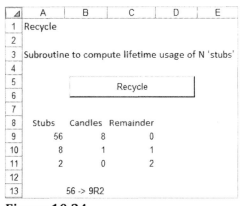

Figure 10.24

```
Sub Recycler()
    Worksheets("Recycle").Select
    Cells(9, 1).Select
    Range(Selection, Selection.End(xlDown)).Select
    Range(Selection, Selection.End(xlToRight)).Select
    Selection.ClearContents
    Cells(9, 1).Select
    stubs = InputBox(Prompt:="How many stubs?")
    If stubs = "" Then Exit Sub
    stubs = Val(stubs)  'convert text to number
    .........
End Sub
```

9. Modify the subroutine in Problem 8 to display a user-form that requests both the number of stubs and how many are needed to make a new candle.

10. Your worksheet has a vertical range starting in A2. The size of the range is between 10 and 500. Each cell holds a value in the range 10 to 99. So we might have 50, 45, 65, 60, 70, 75. You want them sorted in an odd way: all the numbers ending with 0 first, then all those ending with 1, and so on. The numbers are to be sorted in increasing order within each group. So the example would give 50, 60, 70, 45, 65, 75. Write a subroutine that takes the numbers in column A and generates the sorted list in column B. There are many sorting algorithms; pick the one that is easiest to code.

Modeling I

In the preceding chapters we have concentrated on Excel features. In this chapter we are more concerned with getting answers to problems. It is very likely that some of the topics will be outside your interest areas, but please read every exercise since a major objective of the chapter is showing how to lay out a worksheet. This is something with which many new users have difficulty.

Exercise 1: Population Model

An ecological niche contains two species: the prey and the predator. A theoretical analysis of the problem has yielded equations for the successive population of the two species:

$$N_{t+1} = (1.0 - b(N_t - 100))N_t - kN_tP_t$$
$$P_{t+1} = qN_tP_t$$

where:

N_t = the population of prey in generation t
b = the net birth-rate factor
k = the kill-rate factor
P_t = the population of predators in generation t
q = efficiency in use of prey.

We wish to observe how this model predicts the changing populations and to examine the sensitivity of the model to the values of the parameters. Figure 11.1 shows the worksheet we will make. The values for the five parameters were taken from an ecology textbook. The N and P values are the populations in a square kilometer.

(a) Open a new workbook. On Sheet 1 enter the data shown in rows 1 through 9 of Figure 11.1. Format the worksheet as shown. It is sometimes useful to document the worksheet as we have in rows 3 and 4. An alternative method is to use *Insert | Text | Object* and select *Microsoft Equation 3,* which we discuss in Chapter 17.

◢	A	B	C	D	E
1	Prey and Predator				
2					
3	Prey	N(t+1) = (1.0-b(N(t) - 100) N(t) - kP(t)			
4	Predator	P(t+1) = qN(t)P(t)			
5					
6	No	Po	b	k	q
7	50	0.2	0.007	0.5	0.021
8					
9	t	N(t)	P(t)		
10	0	50	0.20		
11	1	62.50	0.21		
12	2	72.34	0.28		
13	3	76.38	0.42		
14	4	73.02	0.67		
15	5	62.29	1.03		

Figure 11.1

(b) Select A6:E7 and using *Formulas | Defines Names | Create from Selection* name the cells in row 7.

(c) Fill the range A10 to A110 with the series 0 through 100. One method is to type the first two numbers, select A10:A11 and drag the fill handle down to A110. Another is to enter 0 in A10 and then use *Home | Editing | Fill and select Series*.

(d) In B10 and C10 enter =No and =Po, respectively.

(e) The formulas in B11 and C11 for the next generation are =(1-b*(B10-100))*B10-k*B10*C10 and =q*B10*C10. Copy these down to row 110 by selecting them both and clicking C11's fill handle.

(f) Construct a chart similar to that shown in Figure 11.2. This shows just 40 generations but more should be included to observe the effect of altering the.parameters.

(g) We can now experiment with the parameters. After observing the effect of changing one parameter, reenter its original value before changing the next.
   (i)   Change No to 60, 70, ...
   (ii) Change Po to 0.25, 0.3, ...
   (iii) Change b to 0.0055, 0.0006, 0.00065, ...
   (iv) Change k to 0.25, 0.3, 0.19 ,0.18, ...

Which parameters may be changed slightly while maintaining a stable state? Can you find one parameter for which a small change results in either a population explosion or an extinction?

(h) Save the workbook as Chap11.xlsx.

Exercise 2: Vapor Pressure of Ammonia

We have a table[1] (A3:B12 in Figure 11.2) showing the measured vapor pressure of ammonia at temperatures in the range 20 to 60°C. Our task is to estimate the vapor pressure at 75°C and to calculate the density of ammonia at this temperature from the Clausius-Clapeyron equation, given that the heat of vaporization (ΔH) is 1265 kJ/kg.

	A	B	C	D	E	F	G	H	I
1	Ammonia: vapor pressure and density								
2									
3	Temp (°C)	Pressure (kN/m²)		Calculations					
4	20	805		Power	4	3	2	1	0
5	25	985		Coeff of P	1.31E-04	-0.014	0.688	19.759	223.788
6	30	1170		Coeff of dP/dT		0.001	-0.041	1.377	19.759
7	35	1365							
8	40	1570		Specified T (°C)	75				
9	45	1790		Terms for P	4130.245	-5799.825	3872.378	1481.925	223.788
10	50	2030		Computed P (kN/m²)	3909	Vapor pressure at 75°C			
11	55	2300							
12	60	2610		Terms for dP/dT		220.280	-231.993	103.263	19.759
13				Computed dP/dT	111.309				
14									
15				ΔH	1265	kJ/kg			
16				density ρ	30.6	kg/m³			

Figure 11.2

[1]This problem is from *Mathematical Methods in Chemical Engineering* by Jenson and Jeffreys, Academic Press (2000). These authors use the method of finite differences to get values: vapor pressure. = 3880 kN/m² and density = 29.0 kg/m³.

If we plot this data, we find that a third-order polynomial has an R^2 value of 1.00, but to be safe we will use a fourth-order fit.

(a) On Sheet2 of Chap2.xlsx, enter all the text seen in Figure 11.2 and the values in columns A and B.

(b) The values in E4:I4 are the powers for each LINEST term; enter these for use later. Select E5:I5 and enter =LINEST(B4:B12,A4:A13^{1,2,3,4}) as an array formula. Ignore F6:I6 for now.

In rows 8 through 10 of the Calculations area, we estimate the vapor pressure of ammonia at 75°C.

(c) Enter the specified temperature in E8. We now use the polynomial terms to compute each term in $aT^4 + bT^3 + cT^2 + dT + e$. The formula in E9 is =E5*$E$8^E4 and this is copied across to I9. In E10 we obtain the required vapor pressure by summing the terms with =ROUND(SUM(E9:I9),0).

The Clausius-Clapeyron equation is shown below, where ΔH is the latent heat in kJ/kg, T is in Kelvins and V_L and V_v are the volumes in cubic meters of one kilogram of the liquid and vapor, respectively.

$$\frac{dP}{dT} = \frac{\Delta H}{T(V_V - V_L)}$$

We make the approximation that the volume of the liquid V_L is negligible compared to the volume of the vapor V_v. Then we rearrange the equation to give:

$$\rho = \frac{1}{V_V} = \frac{T}{\Delta H}\left(\frac{dP}{dT}\right)$$

We have a polynomial expression for P, which we may differentiate to get dP/dT. The first term in P is aT^4 giving $4aT^3$ as the first term in dP/dT, and so on. So the coefficient for the first term is $4a$; for the second it is $3b$, and so on.

(d) In F6 enter the first coefficient for dP/dT using =E4*E5. Copy this across to I6 to get the other terms.

(e) Now that we have the coefficients for dP/dT, we may compute each term in $4aT^3 + 3bT^2 + cT + d$. In F12 enter =F6*$F$8^F4 and copy across to I12. Sum these in E13 using =SUM(F12:I12) to give dP/dT.

(f) Enter the given value for ΔH in F15 and compute the density in E16 using =((E8+273)/E15)*E13. Note the use of 273 to convert from Celsius to Kelvin.

(g) Save the workbook.

Exercise 3: Stress Analysis

We will assume every member is in tension so the forces act away from each joint. In the solution, a negative value will indicate compression.

This Exercise will demonstrate a practical use of matrix algebra to solve a system of linear equations. Consider the structure represented by Figure 11.3. We will assume the members of the structure are massless and are freely jointed one to another. Our task is to find the stress in each member. For equilibrium, at each joint the sum of the horizontal and the sum of the vertical components of all forces must be zero. From this we develop the equations shown in the table below the diagram. Next we need to translate this to an Excel worksheet.

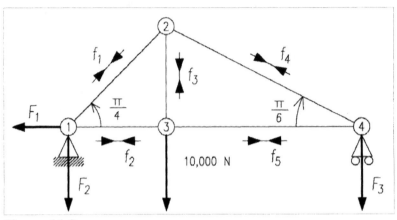

Figure 11.3

Joint	Horizontal	Vertical
1	$-F_1 + \frac{\sqrt{2}}{2} f_1 + f_2 = 0$	$\frac{\sqrt{2}}{2} f_1 - F_2 = 0$
2	$-\frac{\sqrt{2}}{2} f_1 + \frac{\sqrt{3}}{2} f_4 = 0$	$-\frac{\sqrt{2}}{2} f_1 - f_3 - \frac{1}{2} f_4 = 0$
3	$-f_2 + f_5 = 0$	$f_3 = 10000$
4	$-\frac{\sqrt{3}}{2} f_4 - f_5 = 0$	$\frac{1}{2} f_4 - F_3 = 0$

(a) On Sheet3 of Chap11.xlsx enter the text shown in Figure 11.4. Do not bother with the explanatory text boxes.

(b) In A4:H11 enter the coefficient for the equation as shown in the table using the formulas =SQRT(2)/2, =SQRT(3)/2, etc. as appropriate.

(c) In J15:J22 enter the constants for each equation; all but one are zero.

(d) In A15:H22, use MINVERSE to compute the inverse of the coefficient matrix. And in K4:K11use MMULT to multiply the inverse matrix and the constants matrix to give the solutions to the problem.

	A	B	C	D	E	F	G	H	I	J	K
1	Structural Analysis									Double check	0.000000
2	F1	F2	F3	f1	f2	f3	f4	f5			
3			Matrix of equation coefficients							forces	Solution
4	-1	0	0	0.707	1	0	0	0		F1	0.00
5	0	-1	0	0.707	0	0	0	0		F2	-6339.75
6	0	0	0	-0.707	0	0	0.866	0		F3	-3660.25
7	0	0	0	-0.707	0	-1	-0.5	0		f1	-8965.75
8	0	0	0	0	-1	0	0	1		f2	6339.75
9	0	0	0	0	0	1	0	0		f3	10000.00
10	0	0	-1	0	0	0	0.5	0		f4	-7320.51
11	0	0	0	0	0	0	-0.866	-1		f5	6339.75
12											
13											
14			Inverse of Matrix							Constants	
15	-1.00	0.00	-1.00	0.00	-1.00	0.00	0.00	-1.00		0	
16	0.00	-1.00	-0.37	-0.63	0.00	-0.63	0.00	0.00		0	
17	0.00	0.00	0.37	-0.37	0.00	-0.37	-1.00	0.00	×	0	
18	0.00	0.00	-0.52	-0.90	0.00	-0.90	0.00	0.00		0	
19	0.00	0.00	-0.63	0.63	-1.00	0.63	0.00	-1.00		0	
20	0.00	0.00	0.00	0.00	0.00	1.00	0.00	0.00		10000	
21	0.00	0.00	0.73	-0.73	0.00	-0.73	0.00	0.00		0	
22	0.00	0.00	-0.63	0.63	0.00	0.63	0.00	-1.00		0	
23											
24		{=MINVERSE(A4:H11)}						{=MMULT(A15:H22,J15:J22)}			
25											

Figure 11.4

This looks like a simple problem, but the book from which it is adapted made a simple mistake. The author had the wrong sign before ½ in the equation for the vertical components of joint 2. The matrix algebra still gave an answer—the wrong answer! Always look to see if there is another piece of information that can be used to double check the result. In this case we know that the sum of the external forces F_1 and F_2 must balance the 10,000N mass.

(e) Write a formula in K1 that checks this. Use conditional formatting to color H1:K1 red if there is an imbalance; otherwise have them colored green.

(f) Save the workbook.

Exercise 4: Circuit Analysis

In this Exercise we again use matrix algebra. Figure 11.5 shows a circuit with three "meshes." We are asked to find the current in each mesh given that V_s=120V, R_1=2Ω, R_2=5Ω, R_3=2Ω, R_4=4Ω, R_5=15Ω, and R_6=5Ω. We will assume the current in each mesh flows clockwise. We shall apply Kirchhoff's voltage law, which states the algebraic sum of the voltages around a closed loop in a circuit is zero. This gives us three equations:

$$V_s - I_1 R_1 - (I_1 - I_2)R_2 - I_1 R_3 = 0$$
$$-(I_2 - I_3)R_4 - (I_2 - I_1)R_2 = 0$$
$$-I_3 R_5 - I_3 R_6 - (I_3 - I_2)R_4 = 0$$

Substituting the known values and rearranging the equations in preparation for the matrix solution, we write:

$$9I_1 - 5I_5 + 0I_3 = 120$$
$$-5I_1 + 9I_2 - 4I_3 = 0$$
$$0I_1 - 4I_2 + 24I_3 = 0$$

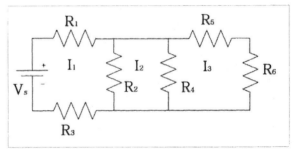

Figure 11.5

◢	A	B	C	D	E	F
1	Kirchhoff's Law					
2						
3		Coefficients			Constants	Solutions
4	I1	I2	I3			
5	9	-5	0	120	I1	20
6	-5	9	-4	0	I2	12
7	0	-4	24	0	I3	2

Figure 11.6

Does something appear to be missing? Where is the range holding the inverse matrix of the coefficients? Read on.

(a) Open Chap11.xlsx and on Sheet 4 begin the worksheet by copying from Figure 11.6 all the text entries and all the values in columns A through D.

(b) Select F5:F7 and enter this array formula: =MMULT(MINVERSE(A5:C7),D5:D7). So that is where the matrix gets hidden—inside the multiplication! The author is not a great fan of nesting complex formulas when it is unnecessary; after all there is lots of space on a worksheet. Also, nesting can be a source of error, but it is fairly safe in this example. Save the workbook.

Exercise 5: Ladder Down the Mine

The purpose of this Exercise is to show that some problems can be solved graphically, especially where limited precision is required. The ladder-in-the-mine is a problem often presented in computing text. Generally, this problem is used as an example of a maximization/minimization problem. We will solve it graphically.

Our ladder is to be taken around a 123° corner where a 9 ft passage joins a 7 ft passage. We assume the "ladder" is actually a piece of machinery that cannot be tilted. It can be shown (see, for example, *Applied Numerical Analysis,* Gerald and Wheatley, Addison-Wesley) that the maximum length of the ladder is found by minimizing the expression:

$$L = \frac{w_1}{\sin(\pi - a - c)} + \frac{w_2}{\sin(c)}$$

where the variables are as shown in the diagram that follows.

For a change of pace, the reader is asked to develop the worksheet on Sheet 5 of Chap11.xlsx and to save the workbook upon completion. Some items to note:

(i) The cells A5:C5 have been named as *a*, *w1_*, and *w2_*, respectively. The underscore was applied by Excel in the naming process since W1 and W2 are valid cell references.

(ii) The cells A5 and B5 were given a custom format so as to display *deg* and *ft*. The *Format Painter* on the *Home / Clipboard* group was used to give other cells these formats. This is quicker than reformatting every cell.

(iii) The formula in B8 is =w1_/SIN(PI()-RADIANS(a)-A8) + w2_/SIN(A8) and is copied to B36. Since the value in the cell named *a* is degrees, we need the RADIANS function here.

(iv) A quick chart was first made with values of c from 0.1 to 0.9; this showed the minimum lies in $0.4 < c < 0.6$. So values from 0.3 to 0.65 in 0.0125 increments were used in the final chart.

(v) On the assumption that the developer of the worksheet might wish to print it for documentation purposes (to show the boss how she got the result): (i) the equation was added with *Insert | Object | Microsoft Equation3*; (ii) Microsoft Visio was used to make the diagram; and (iii) some rows were hidden to give a smaller print out. This last feature means that when making the chart we need to right click it and open the Select Data dialog and using the Hidden and Empty Cells button, specify that hidden rows are to be plotted.

(vi) The formula in D5 is =MIN(B8:B36). To find the value of c that generated the minimum L value, we use in E5 =DEGREES(INDEX(A8:A36,MATCH(D5,B8:B36,0))). The MATCH function finds the row where the minimum is found; INDEX takes the cell in that row from the c column; DEGREES converts the value from radians to degrees.

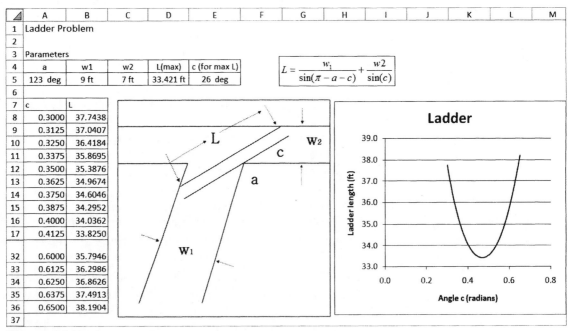

Figure 11.7

This is one of the few presentation-worthy worksheets in the book. In Chapter 17 you will find tips on the Equation Editor and a brief discussion on Microsoft Visio.

Exercise 6: Adding Waves

This brief charting exercise demonstrates how sine waves may be added to generate beats. It is essential to use sufficient points to represent a sine wave. A good rule-of-thumb is to use time increments of $1/(12f)$ where f is the frequency of the sine function as expressed by $Sin(2\pi f)$. Anything much larger than this will fail to generate a sine wave. The reader may wish to confirm this by experimentation.

(a) Working from Figure 11.8, make a worksheet on Sheet 6 of Chap11.xlsx. In B3:C4, we specify the parameter of the two waves to be plotted. In B5 the formula =1/(12*B3) is used to find a time increment. A similar formula is used in C5 while B6 has =MIN(B5:C5) to select the smaller time interval.

(b) A9 starts with $t = 0$ and A10 has =A9+$B$6, and this is copied down to A207.

(c) Cell B9's formula is =B$4*SIN(2*PI()*B$3*$A9). A careful use of mixed references allows us to copy this to C9. Then we can use B9:C9 to fill columns B and C down to row 207.

	A	B	C	D	E	F	G	H	I	J
1	Adding Sine Waves									
2		Wave A	Wave B							
3	Frequency	440	500							
4	Amplitude	1	1.5							
5	delta	0.000189	0.000167							
6	Δt	0.000166667								
7										
8	Time	Wave A	Wave B	Wave C						
9	0.00000	0.00000	0.00000	0.00000						
10	0.00017	0.44464	0.75000	1.19464						
11	0.00033	0.79653	1.29904	2.09557						
12	0.00050	0.98229	1.50000	2.48229						
13	0.00067	0.96316	1.29904	2.26220						
14	0.00083	0.74314	0.75000	1.49314						
15	0.00100	0.36812	0.00000	0.36812						
16	0.00117	-0.08368	-0.75000	-0.83368						
17	0.00133	-0.51803	-1.29904	-1.81707						
18	0.00150	-0.84433	-1.50000	-2.34433						
19	0.00167	-0.99452	-1.29904	-2.29356						

Figure 11.8

(d) Make an XY chart of the data in A9:C207 before adding the formulas in column D. This data series is formatted to show a smooth line with no markers. The two axes are formatted with *None* for the *Major tick marks, Minor tick marks,* and *Tick mark labels*. With a chart with so much data, it can be useful to format the data series setting the *Line Style Width* to a small number.

We next create the data for the superimposition of the two waves and plot it.

(e) The formula in D9 is simply =B9+C9. This is copied down the column.

(f) Select A9:A207, hold down Ctrl, and select D9:D207. Use the *Insert | Chart* command to make the second chart; this is also a smooth XY chart.

(g) Save the workbook

The lower chart clearly shows the development of beats when two sound waves of similar frequency interfere. The techniques of this exercise may be expanded to experiment with three or more waves.

Exercise 7: Centroid of a Polygon

In this Exercise we see an example of a UDF that solves a real-world problem. We wish to enter the coordinates of the vertices of a polygon into a worksheet and have the UDF return the area of the polygon and the coordinates of its centroid. We shall, of course, restrict ourselves to plates of uniform thickness and density.

Let the polygon have N sides (and N vertices). The formulas we shall use are:

$$A = \tfrac{1}{2}\sum_{i=0}^{N-1}(x_i y_{i+1} - x_{i+1}y_i)$$

$$C_x = \tfrac{1}{6A}\sum_{i=0}^{N-1}(x_i + x_{i+1})(x_i y_{i+1} - x_{i+1}y_i)$$

$$C_y = \tfrac{1}{6A}\sum_{i=0}^{N-1}(y_i + y_{i+1})(x_i y_{i+1} - x_{i+1}y_i)$$

These equations are applicable for non–self-intersecting polygons (no line in the polygon's drawing crosses another); this is not a serious constraint, as we are interested in polygons that can be made into physical objects.

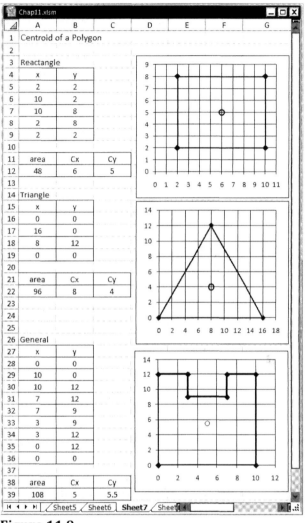

Figure 11.9

The summations in our equations go from zero to N-1, so there will be N terms in each. This shows that we must include the first vertex at the start and at the end of our list of coordinates.

If the formulas are used going counterclockwise around the polygon, then positive values result; otherwise they are negative. We shall avoid this by the use of the *Abs* function in the macro.

(a) Open Chap11.xlsx. Review Chapter 9 if necessary and enter this function on a new module sheet.

```
Function Centroid(myrange)
   Dim temparray(3)
   Set mydata = myrange
   mylast = mydata.Count / 2
   area = 0      'area
   cx = 0        'centroid x-value
   cy = 0        'centroid y-value
   For j = 1 To mylast - 1
   xj = mydata(j, 1)
   xk = mydata(j + 1, 1)
   yj = mydata(j, 2)
   yk = mydata(j + 1, 2)
    term = (xj * yk - xk * yj)
   area = area + term
   cx = cx + (xj + xk) * term
   cy = cy + (yj + yk) * term
   Next j
   area = Abs(area / 2)
   cx = Abs(cx / (6 * area))
   cy = Abs(cy / (6 * area))
   temparray(0) = area
   temparray(1) = cx
   temparray(2) = cy
   Centroid = temparray
End Function
```

Some features to note are: (i) The input range is passed to a VBA object with the *Set* statement allowing us to use subscripts as in *xj=mydata*(j,1); (ii) the use of the *term* variable to save computing the same quantity three times; and (iii) the use of the array *temparray* so that we can return three values to the cells with our function. This also requires the function to be entered as an array function.

(b) Referring to Figure 11.9, the range A5:B9 holds the coordinates for our first test polygon, which is a rectangle. A chart is to be made using these on Sheet7.

(c) Select A12:C12, enter =*Centroid(A5:B9)* and commit the array formula with [Ctrl]+[⇧ Shift]+[↵]. Note that the result is the expected one; this gives us confidence to use the method for complex polygons.

(d) Select and copy B12:C12 then use the Paste Special method to add a new data series to the chart. This plots the centroid.

(e) Complete the worksheet for the next two polygons.

(f) Save the worksheet. Note that, as it now contains a macro, it must be saved as a macro-enabled workbook, and it will be given the extension *.xlsm*.

You may wish to experiment by making changes to the coordinates for the last polygon or by extending the worksheet with further polygons.

Exercise 8: Finding Roots by Iteration

There are many iterative methods of finding the roots of an equation. These include: successive approximations, bisection, secant, and the Newton-Raphson methods. Powerful and instructive as these are we shall look at only one simple example. Chapter 12 introduces the Solver tool, which saves us a great deal of work in locating roots of equations.

Scenario: It cost $P(w)$ dollars to produce w pounds of a chemical where $P(w) = 1000 + 2w + 3w^{2/3}$. The chemical sells for \$4/lb. How much must be sold to break even?

Clearly, the break-even quantity is given by:

$$\begin{aligned} \text{Revenue} &= \text{Expenditure} \\ 4w &= 1000 + 2w + 3w^{2/3} \end{aligned}$$

This tells us we need to solve: $1000 - 2w + 3w^{2/3} = 0$

If we write this as $w = 500 + (3/2)w^{2/3}$ and assume w is small enough that $w^{2/3}$ is insignificant compared to w, then we get our first approximation of $w = 500$.

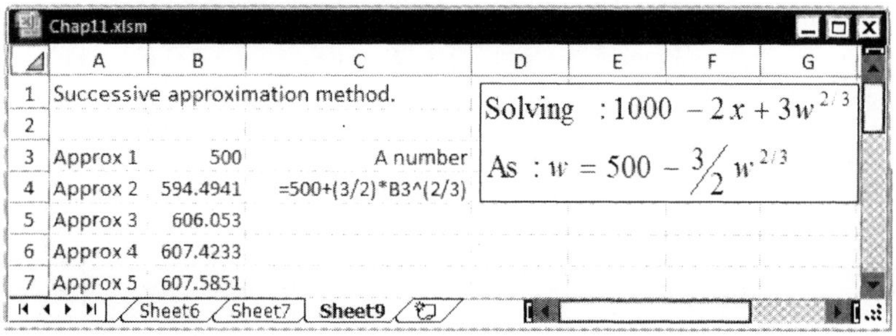

⁴	A	B	C	D	E	F	G
1	Successive approximation method.						
2							
3	Approx 1	500	A number				
4	Approx 2	594.4941	=500+(3/2)*B3^(2/3)				
5	Approx 3	606.053					
6	Approx 4	607.4233					
7	Approx 5	607.5851					

Solving : $1000 - 2x + 3w^{2/3}$

As : $w = 500 - \frac{3}{2}w^{2/3}$

Figure 11.10

(a) On Sheet9 of Chap11.xlsm, enter the text seen in Figure 11.10.

(b) In B3 enter the starting approximation of 500.

We then use this value (500) to compute $500 + (3/2)w^{2/3}$ as the next approximation.

(c) In B4 use the formula =500+(3/2)*B3^(2/3) to compute the next approximation.

(d) Copy this down the column to get the next approximation.

(e) Repeat until the required agreement between successive approximations is obtained. In our case the answer is that 607 pounds must be sold to break even. We just need the successive approximations to have the same integer value.

(f) Save the workbook.

Problems

1. *Using the method of successive approximations (see Exercise 8), find the molar volume of CO_2 at 1 atm and 500K using the van der Waals equation. For CO_2, a = 3.592 $L^2 \cdot atm \cdot mole^{-1}$ and b = 0.04267 L·atm. For your first approximation use the value given by the Ideal Gas Law. Let R have a value of 0.082057 $L \cdot atm \cdot K^{-1} \cdot mol^{-1}$.

2. *Use matrix math to find the forces in each member of the structure shown in Figure 11.11.

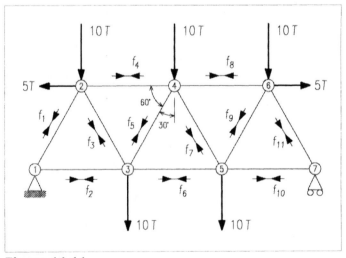

Figure 11.11

3. Figure 11.12 shows an electrical network. Assume currents I_1, I_2 and I_3 are running clockwise around the three loops. Ohm's Law relates the current to the voltage drop at each resistor. Use Kirchhoff's Voltage Law (sum of voltages in a closed loop is zero) to find the three currents. Then use Kirchhoff's Current Law (sum of currents flowing into a node is zero) to find the voltage at each node. How well do the two sets of results agree?

$$I_{ij} = \frac{V_i - V_j}{R_{ij}}$$

For each loop $\sum (\text{pd at each resistor}) = 0$

At each node $\sum (\text{current flowing in}) = 0$

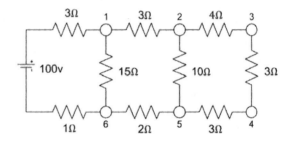

Figure 11.12

4. This "nonscientific" problem was found in a newspaper puzzle section but makes for an interesting programming exercise. You throw 100 coins on a table randomly. Then you randomly pick one coin. If it is a head, you turn it over. If it is a tail, you spin it. What will be the final distribution of head to tails? Your task is to write a subroutine to generate the numerical data shown in Figure 11.13 for N 0 to 100,000.

Can you develop an analytical argument that explains your findings in Problem 4?

◢	A	B	C	D	E	F	G
1	Coins						
2							
3	N	Head	Tails	H/T			
4	0	42	58	0.72		CoinSpin	
5	5000	29	71	0.41			
6	10000	35	65	0.54			
7	15000	29	71	0.41			
8	20000	32	68	0.47		Average H/T	
9	25000	32	68	0.47		0.51	
10	30000	29	71	0.41			
11	35000	24	76	0.32			
12	40000	29	71	0.41			

Figure 11.13

(a) Use the code shown in Problem 8 of Chapter 10 as a start.

(b) Write notes on major topics to be thought about. These include: how to simulate the start, how to simulate picking a coin at random, how to print out after every 5000.

(c) Write an algorithm for the main part of the code.

(d) Write the code in stages. Start with getting row 4 to print. Next get it to print a few more rows and check the results. Then complete the code and run it several times.

5. A researcher is studying a system in which there are two competing processes. He wants to draw a trendline through the several points at the start of the data and another through several points toward the end; see Figure 11.14. Your worksheet should allow him to specify the *x*-value cell to use for last point in the first trendline and that for the first *x*-value in the second trendline.

Hints: In Figure 11.14 no trendlines were inserted; the dotted lines are from two supplementary data series. Note that slopes and intercepts were computed based on the *X*-values but additional points were plotted. The functions MATCH and INDIRECT will be helpful.

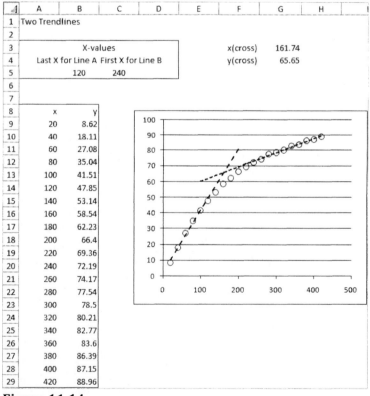

Figure 11.14

Using Solver

Excel has two tools, Goal Seek and Solver, that can save a great deal of time with complex mathematics. From a practical point of view the simple tool Goal Seek is redundant. It has limited scope and is far outpaced by Solver. So why is it there? Simply because, being easier to use, it is less intimidating for the mathematically challenged. So we shall spend a brief time on it.

Solver, which is leased by Microsoft from Frontline Systems Inc., was developed primarily for solving optimization (maximum and minimum) problems. However, it can also be used to solve equations, and that is where we shall start. You may wish to visit www.solver.com to learn more about this product and its variations. The site also has a tutorial for using the Excel Solver, but it concentrates on optimization problems; we will do more with Solver.

In this chapter we will see examples where Solver is used (i) for equation solving, (ii) for curve fitting or regression analysis, and (iii) some simple optimization problems.

Solver needs to be installed on your computer and loaded into Excel. It is likely this happened when Excel was installed. To check, open the *Data* tab and look in the *Analysis* group for a *Solver* icon. If you do not see it, use the Excel Help with the search word *solver* to get instructions on loading it. The Excel Help has nothing more about Solver, as Solver has its own Help facility.

Exercise 1: Goal Seek   Suppose you have an equation such as $\text{Exp}(-x) - \text{Sin}(x) = 0$ and you know (perhaps from making a simple plot) that this has a root such that $0 <= x <= 1$. You could set up a worksheet similar to Figure 12.1 (please ignore the Goal Seek dialogs for now), and by altering the value in A5 and watching B5 you could find what value of x makes the function zero. Think for a moment of what strategy you would adopt. You could confirm that there was a root within (0,1) by making A5 first 0 then 1 and observing that f(x) changes sign. Next, you might next try the midpoint 0.5 and then 0.6. As the sign and magnitude of B5 change, you would modify the direction and amount by which you altered A5 until B5 was nearly 0—or you got tired of the game! Well, Goal Seek works the same way. Let us see Goal Seek at work.

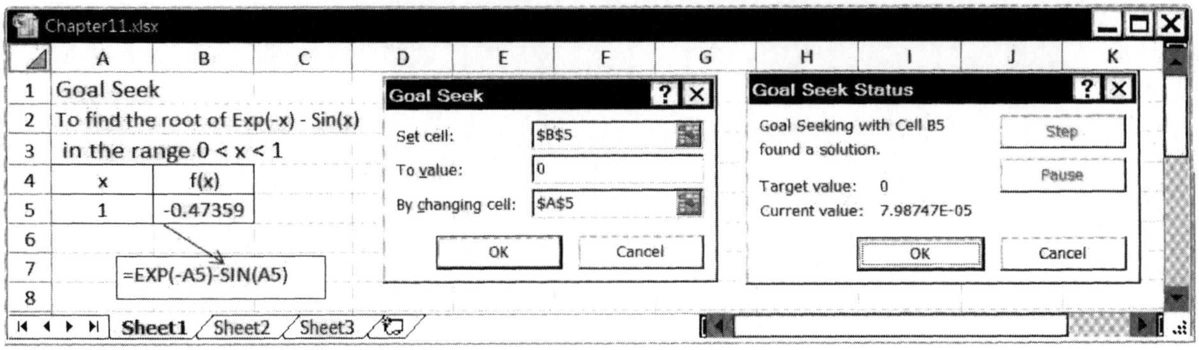

Figure 12.1

(a) On Sheet1 of a new workbook, enter what you see in Figure 12.1 without the text box. The formula in B5 is =EXP(-A5)-SIN(A5).

(b) Use the command *Data | Data Tools | What-If Analysis* to open the Goal Seek dialog.

Note that the *To Value* must be a number; it cannot be a cell reference. So if you want D5 to equal D6, then you will need a cell with =D5-D6, and this will be your *Set Cell* with 0 as the *To Value*.

(c) Our formula is in B5, so this is the *Set Cell*. We want to make this 0, so that is what we type in the *To Value* box, The variable is the value in A5, so this is the *By Changing Cell*. When these have been entered, click the OK button.

(d) Goal Seek now displays its Status dialog giving you the option to either accept what it has found or cancel the operation. Click OK.

(e) Repeat steps (c) and (d) using different starting values (say 0, 1, and 0.5). Note how you have to reenter the problem each time you call up Goal Seek.

Note that the results vary slightly. Goal Seek quits when it has made a certain number of trials (iterations), when a certain time period has passed, or when two answers are within a certain range of each other (convergence limit). There is no way of changing these settings.

(f) If your starting value is 2, Goal Seek will find another root. Make a quick plot and see if you understand why.

(g) Save the workbook as Chap12.xlsx.

Exercise 2: Solver as Root Finder

As an introduction to Solver, we will use it to solve the same problem as in Exercise 1.

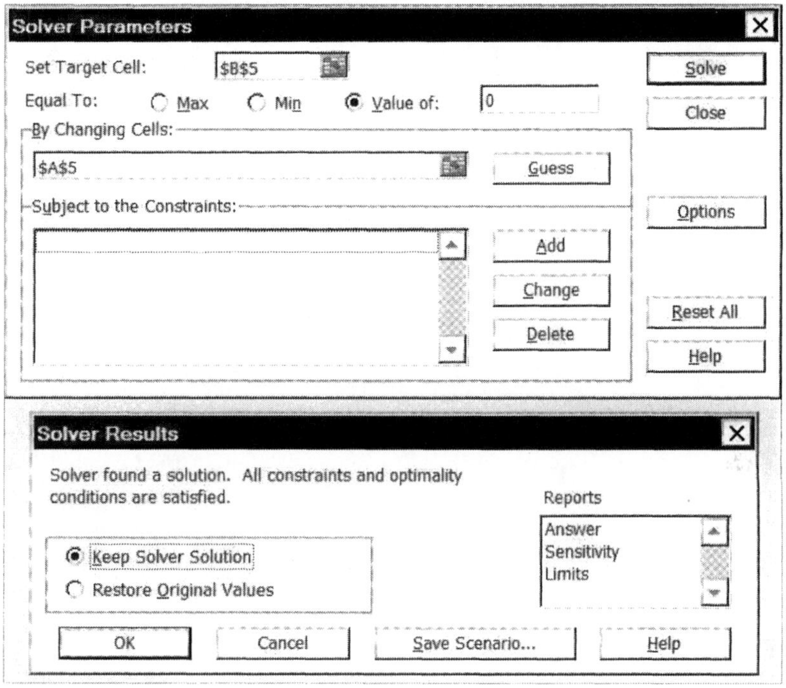

Figure 12.2

(a) Open Sheet1 of Chap12.xlsx. In A5 enter the value 1.

(b) Use the command *Data | Analysis | Solver* to open the dialog shown in the top of Figure 12.2. You can see this is much more detailed than Goal Seek.

(c) For this problem: *The Set Target* is B5, with *Value Of* selected and set to 0, and *By Changing Cells* is A5. Click the *Solve* button in the top right corner.

(d) Solver finds an answer, and the Results dialog pops up. Note that you can accept the answer or return to the original values. The reports are relevant only for optimization problem, not for a Value Of problem. Click OK.

(e) Set A5 to 0 and try again. Note (i) that Solver remembers the problem and (ii) the results are more consistent.

(f) Open Solver again but before you click *Solve*, open the *Options* dialog; see Figure 12.3. We shall not make any adjustments

Note that Solver has its own Help feature.

for this problem, but you may wish to use Solver's Help to learn a little about the first few optional settings.

(g) Save the workbook.

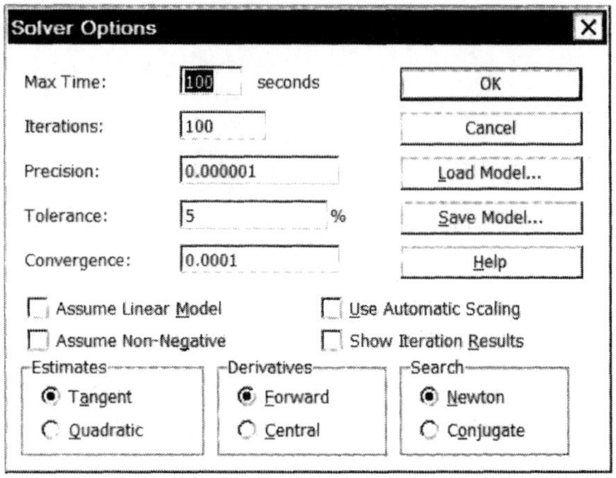

Figure 12.3

Solving Equations with Constraints

THIS IS IMPORTANT: Many Excel texts do not use Solver correctly. It is *not* necessary to have a Set Target Cell. We can find roots of equations using Changing Cells and Constraints. This method can make the problem easier to set up and will often give better results.

If, in the last Exercise, you did look at Solver's Help and read about the Precision setting, you saw that it said: *Controls the precision of solutions by using the number you enter to determine whether the value of a constraint cell meets a target or satisfies a lower or upper bound.* This might sound irrelevant to the problem, but it is not.

With a starting value of 1 in A5 and the default value of Precision at 0.000001 (that's five zeros after the decimal), Solver's answer made B5 a value of $-2.1E-08$ (your result could differ slightly). But when Precision is set to 0.00000000001 (ten zeros after the decimal), the value was $1.8E-12$. Higher precision leads to an answer closer to zero.

This is because: **when the *Value Of* model is used in Solver, it is treated as a constraint problem**. Indeed, to solve the last problem, we could have cleared the *Set Target Cell* and enter a constraint in the Form $B$5 = 0.

This becomes very important when you want more than one cell to take on a certain value. Suppose your model requires all cells in D1:D10 to become zero. If we insist on using the *Value Of* method, we need a single cell, and so write =SUMSQ(D1:D10) in D11 and use it as the *Set Target Cell*. Of course, =SUM(D1:D10) would not work since cells with positive and negative values could sum to zero. A far better way is to use the constraint setting of D1:D10 = 0. This is the method we use in the next exercise.

Exercise 3: Finding Multiple Roots

In solving the simple cubic $x^3+8x^2-9x-72=0$, we will see how the constraints method works and gives superior results. The graph of this function (or simple factorization) shows the roots to be 3, −3, and −8. Solver, like Goal Seek, homes in on the root that is closest to the initial value (sometimes called the *guess*) without passing through a minimum or maximum of the function. So we will be careful with our starting values. In more complex cases, one needs to experiment to find the multiple roots.

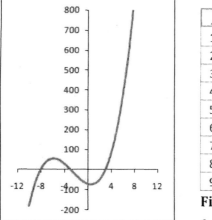

◢	A	B	C	D	E	F
1	Roots of a Cubic Equation with Solver					
2						
3	$f(x) = x^3 + 8x^2 - 9x - 72 = 0$				Models	
4					SUMSQ	CONSTR
5	x1	4	84		FALSE	
6	x2	-4	28		3	3
7	x3	-10	-182		100	FALSE
8						100
9	SUMSQ(C5:C7)	40964				

Figure 12.4

(a) On Sheet2 of Chap12.xlsx, copy from Figure 12.4 the text and values in columns A and B.

(b) The formula in C5 is =B5^3+8*B5^2-9*B5-72, and this is copied down to C7. In C9 we have =SUMSQ(C5:C7). The entries in E:F are for the next exercise.

There is an analytical method for solving cubic equations. This is shown in the workbook CubicEqn.xlsx on the companion website.

We begin by using the "traditional" method of having a cell with a SUMSQ formula as our target cell.

(c) Use Solver as in Exercise 2 with *The Set Target* as *C9*, *Value Of* selected and set to 0, and By Changing Cells as B5:B7. Click the *Solve* button in the top right corner. We get results that are reasonably close to the known roots.

Now we will solve the same problem with no target cell but with a constraint.

(d) Reenter 4, −4, and −8 in B5:B7. Open Solver and clear the *Set Target* box.

(e) In the *Subject to Constraints* area, click the Add button to bring up the Add Constraints dialog; see Figure 12.4. Enter the constraint $C$5:$C$7 = 0. You can type the range reference with or without the $ symbols, or use the pointing method. Click the *OK* button. The *Add* button here is used to add additional constraints.

(f) Use the Solve button to have Solver seek a solution.

(g) Save the workbook.

The results are summarized in the table that follows.

Solver with SUMSQ cell		
Initial	Final	f(x)
4	3.0000015	9.67E-05
-4	-3.0000044	0.000131
-10	-8.0000038	-0.00021

Solver with constraints		
Initial	Final	f(x)
4	3.0000000	0
-4	-3.0000000	0
-10	-8.0000000	1.73E-08

The first two Constraint results are integer values of 4 and -4; the third value is -7.99999999968599. Clearly, the constraint method gave superior results.

Exercise 4: Saving Solver Models

In Exercise 2 we saw that Solver remembers the last used settings. In Exercise 3 we had two Solver models so only the last one used is stored. We can save information that allows us to reconstruct a Solver model.

(a) Return to Sheet2 of Chap12.xlsx and copy from Figure 12.4 the text in E3:F4. The terms SUMSQ and CONSTR are used to remind us of the features of the two models.

(b) Set Solver up to use C9 as the target cell. With E5 as the active cell, open Solver's Option dialog and click on *Save Model*. Solver highlights a 3-by-1 range; click OK. To load the model we shall need to know that three cells were used to solve it. So mark these with borders/and color fills.

(c) Set Solver up to use the constraint and no target cell. With F5 as the active cell, open Solver's Option dialog and click on *Save Model.* Solver highlights a 5-by-1 range; click OK. Mark these with borders/and color fills.

(d) Now you can switch from one model to the next using *Options | Load Model* and selecting with appropriate block of cells.

(e) Save the workbook

Exercise 5: Systems of Nonlinear Equations

In Chapter 4 we saw the use of Excel's matrix functions to solve systems of linear equations. Figures 12.5 and 12.6 show a worksheet and Solver dialog used to solve a system of nonlinear equations. The starting values for *x* and *y* were both 1. The answers are not perfect; they should be integer 2 and 3, but the precision of the method is generally acceptable for real-world problems.

◢	A	B	C	D	E
1	System of Non-linear Equations				
2					
3		$x^2 + 2y^2 = 22$			
4		$-2x^2 + xy - 3y = -11$			
5					
6					
7	x	1.9999999960	F(x,y)	22.0000002927	=x^2 + 2*y^2
8	y	3.0000000257	G(x,y)	-11.0000000057	=-2*x^2 + x*y - 3*y
9					

Figure 12.5

Figure 12.6

Curve Fitting with Solver

In Chapter 7 we used various Excel functions (such as SLOPE, INTERCEPT, LINEST, and LOGEST) to fit experimental data to various mathematical models (linear, polynomial, exponential, etc.). We saw that the theory behind these fitting functions was based on the principle of minimizing the sum of the squares of the residuals. Solver was designed to perform maximization and minimization operations, and so lends itself to curve-fitting problems.

To demonstrate this method, we will do a simple linear fit with some test data taken from the NIST website (www.nist.gov). NIST offers many data sets, together with their fitting parameters to enable others to test their regression programs. We shall use the Norris data set. Figure 12.7 shows a worksheet used to fit the Norris data to $y = mx + b$. The Norris data set is shown in Figure 12.8.

◢	A	B	C
1	Least Squares fit using Solver		
2			
3	SSR	26.61739853	
4		m	b
5	Solver	1.002116799521470	-0.262323073500243
6	LINEST	1.002116818020450	-0.262323073774041
7	NIST	1.002116818020450	-0.262323073774029
8	Solver error	-1.84989821239E-08	2.73785716320E-10
9	Linest error	4.44089209850E-15	-1.23789867246E-14

Figure 12.7

The heading *x*, *y*, and *yfit* in the data set were used to name the columns of data. Cells B5 and C5 were named as *m* and *b*, respectively. The formula in each cell in *yfit* is =mx+b. Note how Excel lets us use *x* to refer to just a single cell in this formula. The formula In B3 is =SUMXMY2(y,yfit). This function conveniently generates the sum of the squares of the residuals. Cells B6:C6 have the formula =LINEST(y,x), while B7:C7 are copies of the values from the NIST website.

With initial values of *m* and *b* as 1, the Solver model used *The Set Target* is *C5*, with *Min box* selected and *By Changing Cells* as **B5:C5**. While the LINEST results are much closer to the accepted NIST values, the Solver answers are quite acceptable.

	E	F	G
1		Norris Data Set	
2	x	y	yfit
3	0.2	0.1	-0.1
4	0.3	0.3	0.0
5	0.3	0.6	0.0
6	0.4	0.3	0.1
7	0.5	0.2	0.2
8	0.6	0.1	0.3
9	10.1	9.2	9.9
10	11.1	10.2	10.9
11	11.6	10.8	11.4
12	118.2	118.1	118.2
13	118.3	117.6	118.3
14	120.2	119.6	120.2
15	226.5	228.1	226.7
16	228.1	228.3	228.3
17	229.2	228.9	229.4
18	337.4	338.8	337.9
19	338.0	339.3	338.5
20	339.1	339.3	339.6
21	447.5	448.9	448.2
22	448.6	449.1	449.3
23	448.9	449.2	449.6
24	556.0	557.7	556.9
25	556.8	557.6	557.7
26	558.2	559.2	559.1
27	666.3	668.5	667.4
28	666.9	668.8	668.0
29	669.1	668.4	670.3
30	775.5	778.1	776.9
31	777.0	778.9	778.4
32	779.0	778.9	780.4
33	884.6	888	886.2
34	887.2	888	888.8
35	887.6	888.8	889.2
36	995.8	998	997.6
37	996.3	998.5	998.1
38	999.0	998.5	1000.9

Figure 12.8

	A	B	C
1	Gaussian Fit		
2			
3	Before		After
4	1600	h	1580.69
5	0.255	mu	0.253959
6	0.005	sig	0.003654
7	0	base	40.11852
8	SSR		114867.2
9			
10	x	y	yfit
11	0.239	25	40.12
12	0.240	24	40.12
13	0.241	39	40.12
14	0.242	49	40.15
15	0.243	56	40.31
16	0.244	84	41.06
17	0.245	66	44.00
18	0.246	97	53.89
19	0.247	158	82.19
20	0.248	244	150.81
21	0.249	353	290.84
22	0.250	444	528.99
23	0.251	773	860.77
24	0.252	1196	1226.11
25	0.253	1677	1515.68
26	0.254	1654	1620.61
27	0.255	1341	1497.53
28	0.256	1173	1197.10
29	0.257	933	830.85
30	0.258	550	505.37
31	0.259	220	275.79
32	0.260	101	142.89
33	0.261	97	78.70
34	0.262	39	52.59
35	0.263	26	43.59
36	0.264	11	40.95
37	0.265	16	40.29
38	0.266	10	40.15
39	0.267	13	40.12
40	0.268	8	40.12
41	0.269	5	40.12

Figure 12.9

Exercise 6: Gaussian Curve Fit

Having demonstrated that this is a viable method of performing regression analysis, we will use it in some more challenging examples

Figure 12.9 shows, in A10:B41, some experimental data that is to be fitted to a Gaussian curve. The function is given by:

$$y_i = h \exp\left(-\left(\frac{x_i - \mu}{\sigma}\right)^2\right) - b$$

where:

y_i = the predicted value
h = the peak height above the baseline
x_i = the value of the independent variable
μ = the position of the maximum
σ = the standard deviation and
b = the baseline offset

(a) On Sheet5 of Chap12.xlsx, start a worksheet similar to that in Figure 12.9. Begin by entering all the text and values except the values in C4.

(b) In C4:C7 use the same values as in A4:A7. Use B4:B7 to name the cells in C4:C7. We are going to vary the C4:C7 cells with Solver but will kept the A4:A7 values to remind us of our starting values.

(c) The formula in C11 is =h*EXP(-(((A11-mu)/sig)^2))+base. There may appear to be an extra pair of parentheses in this, but that is not the case; we need to allow for the fact that the negation operator has the highest priority.

(d) Construct a chart of the data in A10:C41. This will resemble the chart in Figure 12.10 where the markers are the *y*-values and the line the *yfit*-values .

You may have been wondering where the starting values for the *h, mu,* and *sig* parameters came from. The chart will answer this question. The height appears to be about 1600; the midpoint seems to be in the range 0.25 and 0.26 so we use 0.255 for *mu.* The starting value for *sig* is found by experimentation. Try 1 in C6 and see the effect on *yfit.* Now try 0.5 and again see the effect on *yfit.* You will find that 0.005 gets *yfit* to more or less fit the *y*-values. The tails of the curve are not far from zero, so a starting value of 0 for *b* would be appropriate. So now we have reasonable starting parameters.

(e) To get ready for Solver we need a target cell holding the sum of the squares of the residuals. In C8 enter the formula =SUMXMY2(B11:B41,C11:C41).

(f) Use Solver to complete the task. The target cell is *C8*, which we wish to minimize by changing *C4:C7*. The resulting values are shown in Figure 12.9, while Figure 12.10 shows before and after fitting plots.

(g) Save the workbook.

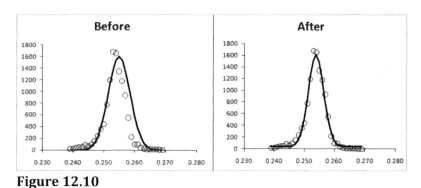

Figure 12.10

Exercise 7: A Minimization Problem

Scenario: An open-top tank is to be made from a sheet of metal by bending and welding (Figure 12.11). The specifications are that the volume is to be 1.0 m^3 using the minimum sheet area. You are to find the dimensions *a* and *b*.

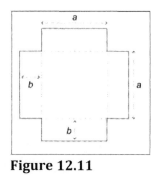

Figure 12.11

The worksheet to solve this problem, together with the Solver dialog, are shown in Figure 12.12.

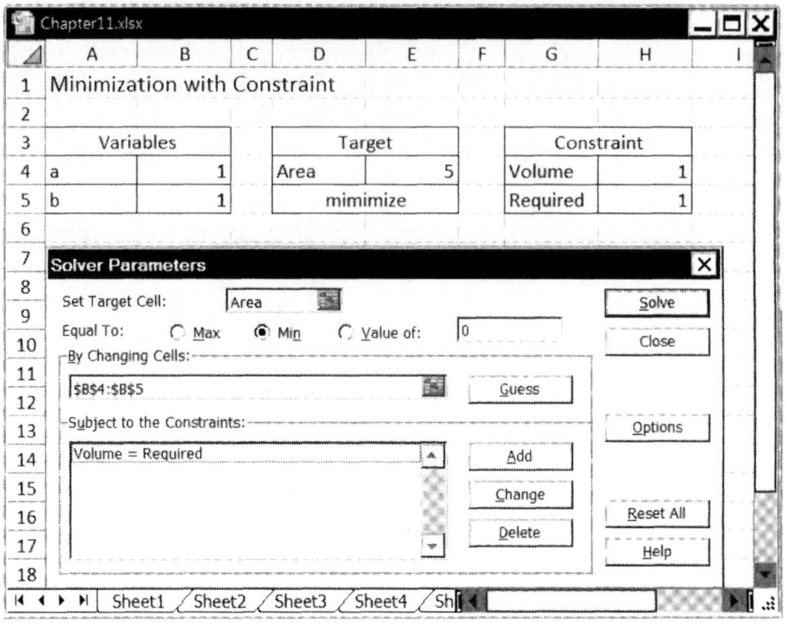

Figure 12.12

(a) On Sheet 7 of Chap12.xlsx, enter the text shown in the figure. Enter the values in B4, B5, and H5.

(b) Select A4:B5, then while holding ⌃Ctrl down, select D4:E4 and G4:H5. Use *Formula | Defined Names | Create from Selection* to name the cells to the right of each text entry.

(c) Enter these formulas: in E4 =(a^2) + (4*a*b) and in H4 =a^2*b. The parentheses in the first formula are just to improve its readability.

(d) Set Solver up as shown in the figure and press the *Solve* button. The results should be 1.26 for *a* and 0.63 for *b*.

(e) Save the workbook.

Exercise 8: An Optimization Problem

Sandbagger, Inc., processes sand to make semi-pure silica to sell to computer chip manufacturers at $50/ton. The company has Plant A and Plant B, in different locations. Plant A can process 450 tons/day at a cost of $25/ton, while Plant B does 550 tons/day for $20/ton.

There are three suppliers: Alpha, Beta, and Gamma. Today, Alpha has 200 tons of sand; they want $10/ton plus shipping of $2/ton to Plant A or $2.50/ton to Plant B. Beta's figures are 300 tons at $9 plus $1 or $1.50 while Gamma has 400 tons at $8 plus $5 or $3 for shipping. Develop a business plan for Sandbagger's operation today.

This is a typical Solver optimization problem. There are three groups of data to be processed; (i) the constants, (ii) the independent variables (called the *decision variables*), and (iii) the dependent variables leading to a problem objective function subject to some constraints. Our constants relate to the two plants and the three suppliers. The independent variables are how much sand from each supplier goes to each plant. The dependent variables are the expenses and income, with the profit being the objective function. The constraints are the finite amount each supplier has and the processing limit of each plant.

With this in mind we plan a worksheet with different areas for the three groups of data. The constraints are placed in the Solver dialog. Figure 12.13 shows our final worksheet.

(a) On Sheet 8 of Chap12.xlsx enter all the text shown in the figure. Enter the values shown in columns B and C.

◢	A	B	C	D	E	F	G	H	I
1	Sandbaggers, Inc								
2									
3		**Parameters**					**Market Plan**		
4	**Selling price**					Tons at each plant			
5	Sand	50	/ton				Plant A	Plant B	Total
6						Alpha	50	50	100
7	**Has two plants**					Beta	50	50	100
8		Plant A	Plant B			Gamma	50	50	100
9	Capacity	450	550	tons		Total	150	150	300
10	Op costs	25	20	$/ton					
11						Expenses			
12	**Can make these purchases**						Plant A	Plant B	Total
13		tons	cost			Alpha	$600	$625	$1,225
14	Alpha	200	10			Beta	$500	$525	$1,025
15	Beta	300	9			Gamma	$650	$550	$1,200
16	Gamma	400	8			Operational	$3,750	$3,000	$6,750
17						Total	$5,500	$4,700	$10,200
18	**Shipping costs**								
19		Plant A	Plant B			**Income**	$15,000		
20	Alpha	2	2.5						
21	Beta	1	1.5			**Profit**	$4,800	To be maximized	
22	Gamma	5	3						

Figure 12.13

(b) Enter the values of 50 into G6:H8 as our starting values for Solver to work with. These are summed in row 9 and column I with formulas such as =SUM(G6:G8).

(c) In G13 enter =G6*($C14+B20) and copy this across and down to fill G13:H16. Sum these values in column I and row 16. Clearly, I17 gives the total of all expenses.

(d) Enter in G19 =I9*B5 (total income) and in G21 =G19-I17 (profit). This last item is our *objective function*.

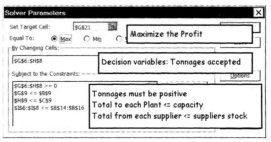

Figure 12.14

(e) All that remains is to run Solver; the settings are shown in Figure 12.14. The maximized profit comes out as $15,375 using all available supplies. Save the workbook.

TK Solver™

One way to test results from a computer program is to set up the same problem in two applications. Some programmers use Excel to test results from a C# program. For the test to be valid, you must totally rethink the problem. It is no good just programming the same algorithm into the second application. We want to test both the algorithm and its implimentation in the computer application.

A number of the problems in this book have been reworked in TK Solver. For many problems, this can be a delightfully easy application to use: you enter *rules* on one sheet and *variables* on the other. Then you tell TK Solver to find the unknown variable. Visit the author's website at people.stfx.ca/bliengme/TKSolver to locate the files and a link for a free trial of TK Solver. This application, which is used in many Engineering schools, can also be interfaced with Excel to expand its capabilities.

Problems

1. Download test data from www.nist.gov for a Gaussian fit. Use the method in Exercise 6 to fit the data. How do your values compare with the accepted values?

2. Redo Exercise 3 of Chapter 11 using Solver.

3. *The vapor pressure ($p°$ in torr) of a pure liquid as a function of the absolute temperature can be expressed as: $\log_{10}(p°) = a - b/T$. The total vapor pressure of a three-component mixture is given by: $P = x_1 p_1° + x_2 p_2° + x_3 p_3°$ where x_i is the mole fraction of component i. Set up a worksheet to use with Solver to find the normal boiling point (the temperature at which P = 760 torr) of a three component mixture. For a working example, use the following data.

	a	b	x
Benzene	7.84125	1750	0.5
Toluene	8.08840	1985	0.3
Ethyl Benzene	8.11404	2129	0.2

4. *A chemical plant[1] uses the following procedure: A volume V ft^3 of solution of compound A having a concentration of a_o lb moles/ft^3 is allowed to react for t_r hours; the vat is then emptied, cleaned and recharged for another cycle. It can be shown that the yield per unit time is given by:

$$yield = \frac{Va_0(1 - \exp(-kt_r))}{t_r + t_c}$$

Make a worksheet using Solver to find the values of t_r that maximize the yield in each simulation shown in the following table. Test the statement that the value of t_r for maximum yield is the solution to the equation:

$$t_r - \ln(t_r k + t_c k + 1)/k = 0$$

V	10	20	10	10	10
a_0	0.1	0.1	0.2	0.1	0.1
k	1	1	1	0.5	1
t_c	0.5	0.5	0.5	0.5	1

5. *What will be the x, y-coordinates for the top right corner of the rectangle in the accompanying figure such that it touches the curve $3y = 18 - 2x^2$ and the area of the rectangle is maximized?

[1] B. Carnahan and J. O. Wilkes, *Digital Computing and Numerical Methods,* Wiley, New York, 1973 (page 435).

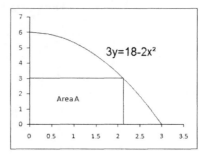

6. *A company has four sources of crude oil.[2] Crude from each source can produce specific amounts of various products. Thus from column B in Figure 12.16 we see that crude A makes 60% gasoline, 20% heating oil, and so on. The company has a market for certain amounts of each product in a week (column G) and fixed supplies from each source (row 10). The profit per barrel is given in row 11. How many barrels of each type should be processed to maximize the profit?

[2]V. G. Jensen and G. V. Jeffreys, *Mathematical Methods in Chemical Engineering* 2nd ed., Academic Press, San Diego, 1977, (page 570).

	A	B	C	D	E	F	G
1	Optimization Problem						
2							
3			Source of Crude Oil				Product
		A	B	C	D	E	orders
4	Product						(bbls)
5	Gasoline	0.6	0.5	0.3	0.4	0.4	170,000
6	Heating Oil	0.2	0.2	0.3	0.3	0.1	85,000
7	Lube Oil	0.0	0.0	0.0	0.0	0.2	20,000
8	Jet Fuel	0.1	0.2	0.3	0.2	0.2	85,000
9	Loss	0.1	0.1	0.1	0.1	0.1	
10	Available (bbls)	100,000	100,000	100,000	100,000	100,000	
11	Profit/bbl	$100	$200	$70	$150	$250	

Figure 12.16

7. *For a change of pace, solve this magazine puzzle. Which three-digit number, when you divide it by the sum of its digits, gives you the sum of its digits plus one?

8. The Langmuir equation relates the amount of gas (S) absorbed on a surface to the pressure (p) of the gas.

$$S = \frac{KS_{max}}{1 + Kp}$$

Fit the data in the following table to find K and S_{max}.

P	4.72	18.45	48.01	79.34	162.31	253.00
S	9.29	18.02	25.08	29.86	38.45	43.48

9. In Problem 15 of Chapter 9, this equation was used to find the surface area of a cylinder with a conical base.

$$S = \frac{2V}{r} + \pi r^2 \left(\csc\theta - \frac{2}{3}\cot\theta \right)$$

We can show by calculus that S is a minimum when the angle of the cone is given by $\theta = \cos^{-1}(2/3)$. Use Solver to confirm that we did the differentiation correctly. How close is Solver's value to the expected one? Can you improve on this?

10. Sutherland's equations can be used to derive the dynamic viscosity of an ideal gas as a function of temperature:

$$\eta = \eta_0 \frac{T_0 + C}{T + C} \left(\frac{T}{T_0} \right)^{3/2}$$

where η is the viscosity (Pa·s) at temperature T, η_0 is the viscosity, T is the input temperature in Kelvin, T_0 is the reference temperature, and C is Sutherland's constant for the specified gas. The following table lists some measured viscosity values for air. Given that for air, $\eta_0 = 18.27 \times 10^{-6}$ Pa·s at 291.15K, find C for air.

t (°C)	10	20	30	40	50	60	70	80	90	100
η (expt)	17.87	18.37	18.86	19.34	19.82	20.29	20.75	21.21	21.66	22.10

11. Refer to Problem 9 in Chapter 2. Make a new worksheet beginning with something similar Figure 12.17. Use Solver to find the *n* values that maximize the profit.

◢	A	B	C	D	E	F	G	H	I	J	K
1	Extraction										
2											
3			Constants								
4	Labor cost		$10	each extraction				A	B	C	D
5	Solvent cost		$0.25	ml			Vs	50	50	50	50
6	Recovery revenue		$50	per g			n	2	0	0	0
7		m0		5	initial mass			m1	1.069	1.069	1.069	1.069
8		Kd		0.43	distribution const						
9		Vw		100	volume of water						

Figure 12.17

Do not use the UDF you may have coded in Problem 7 of Chapter 9 but compute the *m1* values with an Excel function.

13

Numerical Integration

Numerical integration is used to evaluate a definite integral when there is no closed-form expression for the integral or when the explicit function is not known and the data is available in tabular form only. Numerical integration (or *quadrature*) consists of methods to find the approximate area under the graph of the function $f(x)$ between two x-values.

The simplest of these methods uses the *trapezoid rule*. If we divide the area under the curve into a sufficiently large number of parts, as shown in Figure 13.1, then the area under the curve (the approximate integral) is given by:

$$I = \int_a^b f(x)dx \approx \sum_{i=1}^n A_i \tag{13.1}$$

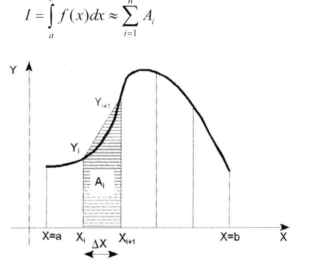

Figure 13.1

We approximate the representative strip to a trapezoid. For a clearer drawing, only five strips are used. Obviously, more, smaller, strips are needed for a good approximation. Let there be n strips and hence $n+1$ data points. The area of a typical strip is given by:

$$A_i = \Delta x \frac{y_i + y_{i+1}}{2} \tag{13.2}$$

Combining the two equations we see that:

$$I \approx \Delta x \frac{y_1 + y_2}{2} + \Delta x \frac{y_2 + y_3}{2} + \cdots + \Delta x \frac{y_n + y_{n+1}}{2} \tag{13.3}$$

Giving:

$$I \approx \frac{\Delta x}{2}\left(y_1 + 2y_2 + 2y_3 + 2y_4 + \cdots + 2y_n + y_{n+1}\right) \tag{13.4}$$

or

$$I \approx \frac{\Delta x}{2}\left(y_1 + 2\sum_{i=2}^{n} y_i + y_{n+1}\right) \tag{13.5}$$

Equation 13.5 is called the *trapezoid rule.* We use the trapezoid approximation in Exercise 1 to evaluate an integral.

A better approximation to the integral is obtained by taking two adjacent strips and joining the three points on the curve with a parabola. This gives Equation 13.6, called the *Simpson ⅓ rule* for approximating the area under a curve.

$$I \approx \frac{1}{3} \sum_{i=1,3,5}^{n=2} (y_i + 4y_{i+1} + y_{i+2})\Delta x \tag{13.6}$$

This rule requires that there be an even number of equally spaced strips. Exercise 2 uses this approximation.

Approximating the curve through *four* adjacent points to a cubic equation gives the *Simpson ⅜ rule* shown in Equation 13.7.

$$I \approx \frac{3}{8} \sum_{i=1,4,7}^{n=3} (y_i + 3y_{i+1} + 3y_{i+2} + y_{i+3})\Delta x \tag{13.7}$$

Surprisingly, the ⅜ rule is often less accurate than the ⅓ rule. However, unlike the ⅓ rule, it does not require an even number of strips. This is an advantage when the data is available only in tabular form (the explicit function being unknown) and there is an even number of data points—an odd number of strips.

Accuracy

If the interval a to b is divided into successively more strips, then, in principle, the accuracy should increase. This is not so in practice. When the number of strips becomes very large, the accumulated round-off error becomes significant.

In some of the exercises and problems, we evaluate definite integrals with known values. There are two reasons for doing this:

we can compare our approximations with the known values to check the accuracy, and it will give us confidence to attack integrals with unknown values.

Exercise 1: The Trapezoid Rule

Using the trapezoid approximation, we evaluate the integral

$$\int_0^{\pi} x\sin(x)dx$$

and show that the approximation yields a result close to the exact result of π. Looking at Figure 13.1 and Equation 13.3, we see that when five strips are used, a total of six $f(x)$ or y-values are needed. In general, when n strips are used then $n+1$ values of y are required. We start this exercise with 10 strips, so we need 11 values of y. For this exercise we will use the trapezoid rule in the form of Equation 13.3. When you have completed this exercise, your worksheet should resemble that in Figure 13.2.

(a) Open a new workbook and enter the text in A1:A6. Enter the following:

    B3: 0   The lower limit of the integration
    B4: =PI()   The upper limit of the integration
    B5: 10   The number of strips
    B6: =(upper-lower)/B5

The last formula computes the value of Δx which we use to find the 11 x-values: *lower, lower* + Δx, *lower* + $2\Delta x$, and so on.

(b) Select A3:B6 and use *Formulas | Defined Names | Create from Selection* to give each cell a name from the text in its neighboring cell to the left.

⌈Ctrl⌉+⌈⇧ Shift⌉+⌈F3⌉ is the short cut for *Formulas | Defined Names | Create...*

(c) Enter the text in row 8.

(d) Enter the following formulas:

    A9: =lower   The value of x_1.
    A10: =A9+delta   The value of x_2

Copy these down to row 19 to give the 11 x-values.

(e) In B9 enter the formula =A9*SIN(A9) to compute the value of y_1. Copy this formula down to B19 giving the 11 y-values for the strips. Note how using the Equation 13.3 form of the trapezoid rule allowed us simply to copy B9 to the other cells.

(f) In C9 enter the formula =delta*(B9+B10)/2 to compute the area of the first strip. Copy this formula down to C18 to get the areas of the other nine strips. Be careful NOT to copy it to C19.

(g) Enter the text in B20:B22.

◢	A	B	C	D
1	Trapezoid Rule			
2				
3	lower	0		
4	upper	3.141593		
5	n	10		
6	delta	0.314159		
7				
8	x	y	strip	
9	0	0	0.015249	
10	0.314159	0.097081	0.073261	
11	0.628319	0.369316	0.177782	
12	0.942478	0.762481	0.307501	
13	1.256637	1.195133	0.434471	
14	1.570796	1.570796	0.528337	
15	1.884956	1.792699	0.56106	
16	2.199115	1.779121	0.511512	
17	2.513274	1.477265	0.369293	
18	2.827433	0.873725	0.137244	
19	3.141593	3.85E-16		
20		Approx	3.115711	
21		Exact	3.141593	
22		Error	-0.82%	

Next to rows 4–5 appears the integral:

$$\int_0^{\pi} x \sin(x)\,dx$$

Figure 13.2

(h) Enter the text in B20:B22. Enter the following formulas:

C20: =SUM(C9:C18)   To sum the 10 trapezoid areas
C21: =PI()   The exact result for the integral
C22: =(C20-C21)/C21   Relative error

(i) Save the workbook as an Excel macro-enabled file called Chap13.xlsm. As we shall be adding modules later in the chapter, we may as well make it macro-enabled right away.

If all has gone well, you will see that the trapezoid rule approximates the integral to 3.115711, which is 0.82% low compared to the exact value. You may wish to modify the worksheet to use 20 strips and note how the error is reduced to 0.21% on the low side.

Exercise 2: Simpson's ⅓ Rule

This exercise uses Simpson's ⅓ rule to find the approximate value of

$$\int_0^2 \exp(x^2)\,dx$$

There is no analytical method for this integral but the published value for the interval [0,2] is 16.452627. Using 20 strips (21 x,y pairs), we shall make a worksheet resembling Figure 13.3. Some rows are hidden to reduce the size of the figure.

◢	A	B	C	D
1	Simpson's ⅓ Rule			
2				
3	lower	0		
4	upper	2	$I = \int_0^2 \exp(x^2)\,dx$	
5	n	20		
6	delta	0.1		
7				
8		x	y	term
9	1	0	1	6.081011
10	2	0.1	1.01005	
11	3	0.2	1.040811	6.591019
12	4	0.3	1.094174	
13	5	0.4	1.173511	7.742942
14	6	0.5	1.284025	
15	7	0.6	1.433329	9.859075
16	8	0.7	1.632316	
25	17	1.6	12.93582	110.442777
26	18	1.7	17.99331	
27	19	1.8	25.53372	227.996083
28	20	1.9	36.96605	
29	21	2	54.59815	
30			Approx	16.455208
31			Published	16.452627
32			Error	0.016%

Figure 13.3

(a) Open the workbook Chap13.xlsm and on Sheet2 enter the text in A1:D8.

(b) Name the cells B3:B6 from the text to their left.

(c) Enter the following values or formulas:
B3:   0   The lower limit of the integration
B4:   2   The upper limit of the integration

B5:   20   The number of strips
B6:   =(upper-lower)/n
The formula in B6 computes the value of Δx, which we use to find the 21 x-values: *lower, lower +Δx, lower +2Δx*, and so on.

(d) The numbers in A9:A29 are not essential but may help you understand how the Simpson rule is implemented in the worksheet. Enter these using the Fill Series method you learned in an earlier chapter.

(e) Enter the initial x- and y-values in B9 and C9:
  B9:   =lower   The value of x_1
  C9:   =EXP(B9^2)   The value of y_1

(f) Enter the second and subsequent values of x and y:
  B10: =B9+delta   The value of x_2
  C10: =EXP(B10^2) The value of y_2
  Copy B10:C10 down to row 29.

When the values of Δx-values are constant, Equation 13.6 becomes:

$$I = \frac{\Delta x}{3} \sum_{i=1,3,5}^{n-2}(y_i + 4y_{i+1} + y_{i+2})$$ (13.8)

Thus we may compute each of the terms, find the summation, and multiply the result by $\Delta x/3$ to approximate the integral.

(g) The first term in the summation is $(y_1 + 4y_2 + y_3)$ so in D9 enter the formula =C9 + 4*C10 + C11. Check that your value agrees with that in Figure 13.3.

(h) We need to copy this formula to each alternate cell in the range D11:D27. The easiest way to do this is select D9:D10 and drag the fill handle down to D27.

The reader is encouraged to extend the worksheet to 1001 data pairs and observe the improved precision.

(i) Enter the text in C30. In D30 enter =ROUND(SUM(D9:D29)*delta/3,6) to find the integral. The rounding is so we can compare with the known value.

(j) Enter appropriate text and formulas in C31:D32 to show the published value and compute the percentage error.

(k) Save the workbook Chap13.xlsm.

Exercise 3: Adding Flexibility

The worksheet in Exercise 2, evaluates a certain integral. For another function, we would need to edit the formula in C9 to reflect the new function to be integrated and copy this down to C29. Another way is to put the function in a module sheet and change the user-defined function each time we wish to evaluate a different integral.

We will begin by solving the same integral to confirm the integrity of the UDF.

The change made in the module will not be reflected in the worksheet until something happens to make Excel recalculate. Pressing F9 will cause this, as will changing the values of the limits.

(a) Open the workbook Chap13.xlsm. We wish to duplicate Sheet2. Hold down the Ctrl key and drag the Sheet1 tab to the right (you will see an icon of a sheet of paper overprinted with a + sign). Release the mouse button and tab labeled Sheet2 (2) will appear. Answer yes to various questions about Named cells; in this way cell B3 of the new sheet will have the name *lower*, and so on.

(b) Use *Home | Editing | Clear* (icon displays an eraser) on C31:D31 as we will be changing functions in this exercise.

(c) Open the Visual Basic Editor with the command *Developer | Code | Visual Basic* or with the Alt+F11 shortcut. Insert a macro sheet of the Chap13.xlsm project and enter this UDF

```
'Function to use with Simpson Rule worksheet
Function SimpFunc(x)
    SimpFunc = Exp(x ^ 2)
End Function
```

Remember: whenever you edit a user-defined function, you must recalculate the worksheet by pressing F9 before any changes in the function will take effect.

(d) Return to the worksheet containing the Simpson rule calculation. Change the formula in C9 to =SimpFunc(B9) and copy this down to C29. Excel will ignore the uppercase letters, but in the Insert Function dialog in the User Defined category our function will show as *SimpFunc*. The values should stay the same as before (see Figure 13.3).

Now we will make a quick change to the UDF to evaluate

$$\int_{-1}^{1} \exp(-x^2)dx$$

(e) Change the third line in the module function to read SimpFunc = Exp(-(x ^ 2)). Carefully note the position of the negation operator relative to the parentheses.

(f) Return to the Simpson worksheet. Change the values of *lower* and *upper* to -1 and 1, respectively. The value of the new function has been calculated. For this function, in the interval -1 to 1, the result should be approximately 1.49.

(g) To make another test, change the user-defined function to Simp = x * Sin(x). Return to the Simpson worksheet and change the values of *lower* and *upper* to 0 and PI(), respectively. How does your result compare to that in Exercise 1?

(h) Save the workbook.

Exercise 4: Going Modular

In this Exercise we implement the Simpson ⅓ rule as a user-defined function in a module. To check our function more easily, we find the value of the same integral as in Exercise 1. In addition, we experiment with making the strip successively smaller and observe how the percentage error changes.

(a) Open the Visual Basic Editor with the [Alt]+[F11] shortcut. On a new module sheet in Chap13.xlsm project, code the integrating function and the function to be integrated as shown in Figure 13.4. Do not type the line numbers; they are for discussion purposes only. The statements in the *Integral* function are examined at the end of the exercise.

(b) On Sheet4 of the Chap13.xlsm workbook enter the text and values shown in A1:A11 of Figure 13.5.

(c) Name the cells B4:B6 with the text to the left of them.

(d) The formulas and values to be entered in B3:B10 are:
   B4:   0   The lower limit
   B5:   =PI()   The upper limit
   B6:   =PI()   The known value of the integral
   B8:   10   The number of strips
   B9:   =Integral(lower,upper,B7)
                          Value from Simpson's rule
   B10: =B8-exact   Error calculation
   B11: =B9/exact   Percentage error

Format B11 to show a percentage value with four decimal places.

```
1   'Simpson One-Third Rule Approximation
2   Function Integral(a, b, n)
3       Integral = 0#
4       delta = (b - a) / n
5       x = a
6       For i = 1 To n Step 2
7           Term = y(x) + 4 * y(x + delta) + y(x + 2 * delta)
8           Integral = Integral + Term
9           x = x + 2 * delta
10      Next i
11      Integral = Integral * delta / 3
12  End Function
13
14  'The function to be integrated
15  Function y(x)
16      y = x * Sin(x)
17  End Function
```

Figure 13.4

◢	A	B	C	D	E	F
1	Simpson's ⅓ Rule					
2						
3	With User-defined Integral Function					
4	lower	0				
5	upper	3.141593	π			
6	exact	3.141593	π			
7						
8	n	10	100	1,000	10,000	100,000
9	Approx	3.141765	3.141593	3.141593	3.141592654	3.141592654
10	Error	0.000172	1.7E-08	1.68E-12	-1.22569E-13	-3.45546E-12
11	% Error	0.0055%	0.0000%	0.0000%	0.0000%	0.0000%

$$\int_0^\pi x \sin(x)\,dx$$

Figure 13.5

(e)   Check that your results in B9:B11 agree with Figure 13.5. If they do not, you may need to edit your module or your formulas.

(f)   Enter the values in C8:F18 and copy B9:B11 across to column F.

(g) Select B7:B10 and drag the handle to column F. Change your *n* values to match those in the spreadsheet. Note that large numbers may be entered with a comma to make them more readable. Do not be surprised if your worksheet takes some time to respond. Microsoft Excel has to do a large number of calculations. Save the workbook.

Note how the absolute error progressively decreases up to *n* = 10,000 and then increases for larger *n* values. Here we are seeing the accumulated round-off errors beginning to creep in.

Explanation of the Integral function:

Line	Comment
3	Type this as **Integral** = 0.0 to tell VBA that it is a real, not an integer, value. This initializes the value of the function to zero.
5	The lower limit of the integral is *a*.
6	We need to sum for odd values of *i* (*i* = 1, 3, 5, ..., *n*). The **step 2** phrase achieves this. It is equivalent to copying the formula in D9 of Exercise 2 to alternate rows.
7	This finds the $(y_i + 4y_{i+1} + yi+_2)$ term for the area of a strip. We multiply by $\Delta x/3$ at the end of the calculation.
8	We keep a running total of the terms computed in line 7. We may read this as: New Integral value = Old Integral value + Term value.
9	This statement computes the *x*-value for the next term.
11	The sum of the partial is multiplied by $\Delta x/3$.

Exercise 5: Tabular Data

There are times when the data to be integrated comes from an experiment and the implicit function is unknown. Which of the three rules should be used to evaluate the integral?

Trapezoid	may be used with any data but is the least accurate.
Simpson ⅓	requires an even number of equally spaced strips.
Simpson ⅜	requires equally spaced *x* values, may be less accurate than the ⅓ rule.

Suppose we have 63 strips (i.e., 64 data pairs). We may use the ⅜ rule for the first three and the ⅓ rule for the remaining 60 strips.

We have seen that increasing the number of strips improves the accuracy of these approximations. With tabular data, this option is not available. While we cannot increase the number of data points since we do not know the function, we can decrease the number by doubling the width of each strip. Essentially, this

Romberg integration can be used to improve the result.

means we ignore every alternate data pair in the table. Obviously, our second value for the integral will be less accurate. This is where Romberg integration is useful. The Romberg integral is computed using:

$$I_R = I_h + \frac{I_h - I_{2h}}{2^n - 1}$$

where:

I_R = the improved value of the approximation
I_h = the value of the approximation with strips of width h
I_{2h} = the value of the approximation with strips of width $2h$
n = 2 for the trapezoid rule and 4 for Simpson's rules

◢	A	B	C	D	E
1	Tabular Data				
2					
3	x	y	h=0.2	h=0.4	
4	1.8	6.050	6.0500	6.0500	
5	2.0	7.389	14.7780		
6	2.2	9.025	18.0500	18.0500	
7	2.4	11.023	22.0460		
8	2.6	13.464	26.9280	26.9280	
9	2.8	16.445	32.8900		
10	3.0	20.086	40.1720	40.1720	
11	3.2	24.533	49.0660		
12	3.4	29.964	29.9640	29.9640	
13					
14			h=0.2	h=0.4	Romberg
15	Integrals		23.99440	24.23280	23.91493
16	Exact		23.91445	23.91445	23.91445
17	Errors		0.334%	1.331%	0.002%

Figure 13.6

In this Exercise we use the trapezoid rule to find an approximation to some tabulated data. Clearly, we may use Romberg integration only when the strips are evenly spaced.

(a) On Sheet5 of the workbook Chap13.xlsm enter the text in A1:D14 as shown in Figure 13.6.

(b) Enter the values in A4:B12. This is the experimental data that we wish to integrate.

(c) We will use the trapezoid rule in the form of Equation 13.4 in this Exercise. Enter these formulas:
C4:   =B4

C5:   =2*B5 and copy this down to C11

C12: =B12

These are the bracketed terms in Equation 13.4. The integral is completed by adding the entry:

C15: =0.2/2*SUM(C4:C12)

(d)   In column D we will use strips of twice the width. Enter the formulas:

D4:   =B4

D6:   =2*B6 and copy this to D8 and D10

D12: =B12

D15: =0.4/2*SUM(D4:D12)

(e)   The Romberg integral is found with

E15: =C15 + (C15 - D15)/3.

(f)   Save the workbook.

Is the Romberg value a better approximation? The data is actually $y = \exp(x)$ with the values rounded to three decimal places. Therefore our result should be the integral

$$\int_{1.8}^{3.4} \exp(x)dx = \exp(3.4) - \exp(1.8) = 23.9144$$

In row 17, compute the percentage errors of the three values. Clearly, the Romberg value is the more accurate one.

Exercise 6: Gaussian Integration

The Gaussian two-point integration formula, as derived in most elementary numerical analysis textbooks, has the wonderful simplicity of:

$$\int_{-1}^{1} f(t)dt = f\left(-\tfrac{1}{\sqrt{3}}\right) + f\left(+\tfrac{1}{\sqrt{3}}\right) \tag{13.9}$$

The four-point formula is only slightly more formidable:

$$\int_{-1}^{1} f(t)dt = \tfrac{5}{9}f\left(-\tfrac{\sqrt{3}}{5}\right) + \tfrac{8}{9}f(0) + \tfrac{5}{9}f\left(+\tfrac{\sqrt{3}}{5}\right) \tag{13.10}$$

These formulas may be generalized to:

$$\int_{-1}^{1} f(t)dt = \sum_{i=1}^{n} w_i f(t_i) \tag{13.11}$$

The accompanying table lists the values for the weights (w_i) and the points (t_i) for various numbers of points in the integration.

The degree of the polynomial function for which each integration formula is accurate is given by $2n-1$. Thus the three-point formula is accurate for polynomials up to degree 5. When one is unsure of the number of points to use, successively use 2, 3, ... points until two results agree to the precision required.

The weights and point values in the table are for the limits of integration ±1. To use them with other limits (a to b), we make the substitution

$$x = \frac{(b-a)t + b + a}{2} \quad \text{so} \quad dx = \left(\frac{b-a}{2}\right)dt$$

giving

$$\int_a^b f(x)dt = \frac{b-a}{2}\int_{-1}^{1} f\left(\frac{(b-a)t+b+a}{2}\right)dt \qquad (13.12)$$

n	$\pm t_i$	w_i
2	$\sqrt{1/3}$	1
3	0	8/9
	$\sqrt{3/5}$	5/9
4	0.33998 10435 84856	0.65214 51548 62546
	0.86113 63115 94052	0.34785 48451 37455
5	0	0.56888 88888 88889
	0.53846 93101 05727	0.47862 86704 99334
	0.90617 98459 38664	0.23692 68850 56189
6	0.23861 91860 83197	0.46791 39345 72691
	0.66120 93864 66264	0.36076 15730 48139
	0.93246 95142 03152	0.17132 44923 79170
8	0.18343 46424 95644	0.36268 37833 78363
	0.52553 24099 16329	0.31370 66458 77887
	0.79666 64774 13017	0.22238 10344 53966
	0.96028 98564 97439	0.10122 85362 90617
10	0.14887 43389 81631	0.29552 42247 14753
	0.43339 53941 29247	0.26926 67193 09997
	0.67940 95682 99032	0.21908 63625 15982
	0.86506 33666 88985	0.14945 13491 50581
	0.97390 65285 17188	0.06667 13443 08648

In this Exercise we use Gaussian integration to find the value of:

$$I = \int_{-1}^{1} x^2 \cos(x)\,dx$$

To make the worksheet more versatile, we will code a user-defined function for $x^2 \cos(x)$; in this way we will be able to perform Gaussian integrations on other functions merely by editing the user-defined function. We will increase the number of terms until we have an approximate value to four decimal places.

(a) Open Chap13.xlsm and invoke the VBE. Insert another module on the Chap13.xlsm project on which to code the following function:

```
Function gaussfunc(x)
    gaussfunc = x ^ 2 * Cos(x)
End Function
```

◢	A	B	C	D	E
1	Gaussian Integration				
2					
3	Terms	Weight	Point	Term	Approx
4	2	1	0.577350269	0.279303943	
5		1	-0.57735027	0.279303943	0.558607885
6					
7	3	0.555555556	0.774596669	0.238234398	
8		0.888888889	0	0	
9		0.555555556	-0.77459667	0.238234398	0.476468795
10					
11	4	0.652145155	0.339981044	0.071064922	
12		0.347854854	0.861136312	0.168076458	
13		0.347854854	-0.86113631	0.168076458	
14		0.652145155	-0.33998104	0.071064922	0.478282759
15					
16	5	0.478628671	0.538469231	0.119140142	
17		0.236926885	0.906179846	0.119993421	
18		0.568888889	0	0	
19		0.236926885	-0.90617985	0.119993421	
20		0.478628671	-0.53846923	0.119140142	0.478267125

Figure 13.7

(b) Return to the workbook and on Sheet6 enter the text shown in A1:E3 of Figure 13.7.

(c) Select the column headings B:E and use the command *Home | Cells | Format | Column Width* to set the width to 14. Or after selecting the headings, right click and open the column width dialog from the shortcut menu.

(d) Enter the values shown in A4:A16 and in B4:B20. It is safer to use =5/9 in B7 and =8/9 in B8 rather than numerical values.

(e) Enter the values shown in C4:C20. For C4 you may wish to use =1/SQRT(3) and the negative of this in C5. Similarly for C7 use =SQRT(3/5) and the negative of this in C9.

(f) In D4 enter =B4*Gaussfunc(C4). Copy this down to D20. Delete D6, D10, and D15.

(g) Move to E5 and click on the Autosum button. Drag over D4:D5 to give the formula =SUM(D4:D5). Repeat this operation for the three other approximations.

(h) Save the workbook.

We can see that the four-point and the five-point approximations agree to four decimal places, so our task is complete.

Exercise 7: Monte Carlo Techniques

There is no mathematical advantage to performing a numerical integration using a Monte Carlo technique. However, it does provide a simple way to illustrate a Monte Carlo calculation. A large worksheet would be needed to model a true stochastic process.

Consider a circle inscribed within a square with sides of l units. The radius (r) of the circle will be $l/2$. A large number (N) of darts are randomly thrown at the diagram, and the number (C) that fall within the circle are counted. If the throwing was truly random, then:

$$\frac{\text{Number of darts in circle}}{\text{Total number of darts}} = \frac{\text{Area of circle}}{\text{Area of square}}$$

or

$$\frac{C}{N} = \frac{\pi r^2}{l^2} = \frac{\pi}{4}$$

Hence the value of π may be approximated from a simple dart-throwing experiment. We will use this Monte Carlo method to get an approximate value of the integral

$$I = \int_0^{10} (-x^3 + 10x^2 + 5x)dx$$

As in previous exercises, we have chosen an integral that can be solved analytically so as to be able to evaluate our result.

⊿	A	B	C	D	E	F	G	H	I
1	Monte Carlo Integration							x	f(x)
2	x	y	in/out		Throws	1000		0	0
3	8.30	92.82	1		Inside	545		1	14
4	0.82	82.35	0		Area	1090.00		2	42
5	7.00	134.98	1		Actual	1083.33		3	78
6	3.86	72.55	1		%error	0.62%		4	116
7	4.87	7.20	1					5	150
8	5.14	135.09	1					6	174
9	1.39	35.47	0					7	182
10	7.92	17.97	1					8	168
11	5.38	169.66	0					9	126
12	2.40	65.26	0					10	50
13	3.72	78.16	1						
14	10.00	6.74	1						
15	8.68	66.90	1						
16	2.02	182.29	0						
17	3.57	144.69	0						
18	7.08	51.51	1						
19	7.42	51.68	1						
20	1.41	91.86	0						
21	2.36	97.92	0						
22	0.85	0.67	1						
23	7.76	99.07	1						
24	6.61	120.46	1						
25	7.49	83.47	1						
26	6.10	126.44	1						
27	7.32	190.88	0						

Figure 13.8

(a) On Sheet7 of Chap13.xlsm begin by constructing the table shown in H1:I12 of Figure 13.8. The formula in I2 is =-H2^3+10*H2^2+5*H2. This is copied down to row 12. Make a chart of the data.

Our curve is enclosed by a 10 by 200 rectangle. We will use the RAND function to generate two random values from which we will find the position of the dart. The function returns a value from 0 to 1, so for the x-value we will use RAND()*10 and for the y-value RAND()*200.

(b) In A3 enter =RAND()*10 and in B3 enter =RAND()*200. Copy these down to row 1002. Your values will not be the same as in the figure.

(c) The formula in C3 is =IF(B3>-A3^3+10*A3^2+5*A3,0,1). This returns 0 when the dart has fallen above the curve, and 1 otherwise. Copy it down to C1002.

(d) The formulas in column F are:

F2:	=COUNT(A3:A1002)	Total darts thrown
F3:	=SUM(C3:C1002)	Darts inside the curve
F4:	=(200*10)*F3/F2	Area under the curve
F5:	=-(1/4)*10^4+(10/3)*10^3+(5/2)*10^2	
		Analytical result
F6:	=(F4-F5)/F5	Relative error

(e) Repeatedly press F9 to recalculate the worksheet. A new set of random numbers is generated each time leading to a new value in F5. The error seems to lie within a range of ±5% of the analytical value. Adding more random numbers would help. Alternatively, we could record the result for, say, 20 recalculations and take an average. This could be done automatically using a VBA subroutine. Save the worksheet.

Problems

1. Estimate $\int_{-1}^{0} \frac{1}{x+2} dx$ with Δx = 0.2 using the trapezoid rule. Try steps of 0.1 to see how much improvement there is.

2. Use Simpson's rule for the integral in Problem 1 with the same steps. Compare the four results with the value obtained by direct integration.

3. Write a UDF to evaluate $\int_{-1}^{0} \sqrt{2x+1} dx$ The header should include an argument *n* to specify how many strips to use.

4. *Find the area bounded by the two functions $f(x) = 4x^2$ and $g(x) = x^4$ in the range 0 <= x <= 2.

5. *Calculate[1] the definite integral

$$\int_{1}^{2} x \ln(x) dx$$

using Simpson's rule with h values of 0.5, 0.25 and 0.125. Given that the exact solution may be found from

$$\int x \ln(x) dx = \left(\frac{x^2}{2}\right) \ln(x) - \frac{x^2}{4}$$

show that the error E(h) in Simpson's rule may be estimated from the inequality:

[1] A. Jeffrey, *Mathematics for Engineers and Scientists*, Chapman and Hall, New York, 1996 (page 770).

$$\frac{nh^5 m}{90} \le E(h) \le \frac{nh^5 M}{90}$$

where n times 2 is the number of intervals, m is $\min|f^{(4)}(x)|$ and M is $\max|f^{(4)}(x)|$ on $a < x < b$. The symbol $f^{(4)}(x)$ stands for the fourth differential of the function being integrated.

6. *With an odd number of data points in Figure 13.9, we cannot use the Simpson ⅓ rule to find the area under the curve. However, we can use the ⅜ rule on the first four points, then the ⅓ rule on groups of three for the rest of the data. What answer do you get?

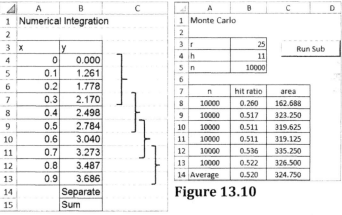

◢	A	B	C
1	Numerical Integration		
2			
3	x	y	
4	0	0.000	
5	0.1	1.261	
6	0.2	1.778	
7	0.3	2.170	
8	0.4	2.498	
9	0.5	2.784	
10	0.6	3.040	
11	0.7	3.273	
12	0.8	3.487	
13	0.9	3.686	
14		Separate	
15		Sum	

Figure 13.9

◢	A	B	C	D
1	Monte Carlo			
2				
3	r		25	Run Sub
4	h		11	
5	n		10000	
6				
7	n	hit ratio	area	
8	10000	0.260	162.688	
9	10000	0.517	323.250	
10	10000	0.511	319.625	
11	10000	0.511	319.125	
12	10000	0.536	335.250	
13	10000	0.522	326.500	
14	Average	0.520	324.750	

Figure 13.10

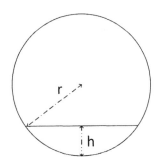

7. Write a subroutine to find the area between the circumference of a circle of radius r and a horizontal line, which is a distance h from the circumference along a radius at right angles to the line. Your subroutine should read the r, h and n values from the worksheet (Figure 13.10), use a function to compute the area, and put data in A8:C14.

8. The equation of a lemniscate is $\rho = a^2\cos(2\theta)$. The length (s) of an arc of its perimeter is given by

$$s = \int_0^\theta \frac{a\,d\theta}{\sqrt{\cos(2\theta)}}$$

Make a worksheet that will find s for $0 \le \theta \le 90$. Make it general enough that it works for any a value.

Challenge! Make this chart in Excel. The author used 100 points. Clearly, at the extremities, closer data points are needed for a smooth plot. This is an ideal project for a VBA subroutine generating the data on the worksheet.

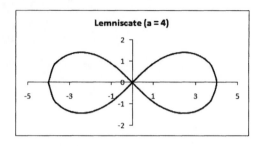

9. *Using either the trapezoid or Simpson method, evaluate the area under the curve represented by the data in the following table. Then plot the data and add a polynomial trendline. Use LINEST to get the coefficients of the polynomial and integrate the polynomial to find the area. How good is the agreement?

x	3.00	3.60	4.00	4.70	5.50	6.25	7.00	7.50
y	1.17	1.02	0.90	0.63	0.59	0.63	0.66	0.79
x	8.60	9.00	9.40	10.00	10.45	10.85	11.25	
y	1.20	1.45	1.65	2.05	2.26	2.62	3.00	

10. Evaluate $\int_{0}^{\pi} \sin(x)dx$ using both Simpson's ⅓ rule and Gaussian integration with $n = 8$. Which method gives the better value?

11. The table that follows gives values for f(x). Using the trapezoid rule, find the integral with steps (h) values of 0.1, 02, and 0.4.

x	1	1.1	1.2	1.3	1.4
f(x)	1.543	1.669	1.811	1.971	2.151
x	1.5	1.6	1.7	1.8	
f(x)	2.352	2.577	2.828	3.107	

12. The values in the table in Problem 11 were generated from f(x) =cosh(x). Show graphically if the error in the numerical integration values are proportional to h.

14

Differential Equations

Differential equations occur in many physical problems. Let us look at some simple examples.

(i) A body falling through the air is subjected to two forces: gravity acting downward and air resistance acting upward. The first force is constant, but the second is proportional to the body's velocity. This gives rise to the first-order differential equation

$$m\frac{dv}{dt} = g - kv^2 \tag{14.1}$$

(ii) Consider the chemical reaction A + B → C where the rate of reaction is proportional to the concentration of A and to the concentration of B. Let x be the amount of A and B reacted at time t, and let the initial concentration of A and B be a and b, respectively. These quantities will be related by

$$\frac{dx}{dt} = k_2(a-x)(b-x) \tag{14.2}$$

(iii) The equation of motion for a harmonic oscillator is

$$\frac{d^2x}{dt^2} + \omega x = 0 \tag{14.3}$$

Equations 14.1 and 14.2 are examples of first-order differential equations, while Equation 14.3 is of second order. The equations in these examples may readily be solved by analytical means. The next equation is also first order, but its solution is rather more difficult to arrive at:

$$\frac{dy}{dx} = \frac{x+y}{x-y} \tag{14.4}$$

When a differential equation is difficult or impossible to solve analytically, we may use numerical methods to find approximate solutions.

Consider the simple equation $dv/dt = g$ for a falling body when air resistance is ignored. This integrates to give $v = gt + c$ where c is the integration constant and g is a constant of known value. Thus we do not have a unique solution, since any value of c will satisfy the differential equation. By inspection of the solution we see that c is the value of v when t equals zero. We need to know this value in order to uniquely solve the equation. In general, to solve $dy/dx = f(x,y)$ over the x range $[a, b]$, we need to know the value of $y(a)$, which is called the *initial value.* Problems of this type are called *initial value* problems. With second-order differential equations two integration constants arise. For an initial value problem we need to know the initial value of the two values of the dependent variables. Alternatively, the problem may be defined by specifying some conditions at one value of x and others at another value of x. Such problems are called *boundary value* problems.

Exercise 1: Euler's Method

Euler developed a method for finding the approximate solution to initial value problems. Let the differential equation to be solved have the form of Equation 14.5 and let the initial value of y be y_0.

$$\frac{dy}{dx} = y' = f(x,y) \tag{14.5}$$

Let the solution (i.e., the integral of Equation 14.5) have the form of Equation 14.6.

$$y = g(x,y) \tag{14.6}$$

Consider the two curves in Figure 14.1 where $f(x)$ is the differential to be solved and $g(x)$ is the integrated solution.

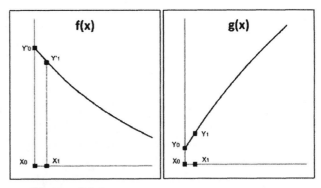

Figure 14.1

From Equation 14.5, we may calculate any value $y'_1, y'_2, ..., y'_n$. We already know the value of y_0 – the initial value. Our task is to find values for $y_1, y_2, ..., y_n$. Integration of Equation 14.5 from x_0 to x_1 yields Equation 14.7.

$$y_1 = y_0 + \int_{x_0}^{x_1} f(x_0, y_0) dx \tag{14.7}$$

The second term on the right is the area under the curve $f(x,y)$ between the two x-values. Euler approximated this to the area of the rectangle defined by $y'_0, y'_1, x_0,$ and x_1. The approximate value of y_1 is then given by Equation 14.8, or by Equation 14.9 when the x increment is represented by h.

$$y_1 = y_0 + (x_1 - x_0)f(x_0, y_0) \tag{14.8}$$

$$y_1 = y_0 + hf(x_0, y_0) \tag{14.9}$$

Having found an approximation for y_1, we may now find an approximation for y_2

$$y_2 = y_1 + hf(x_1, y_1) \tag{14.10}$$

In general, the value of the approximation at one point is found from the previous one using

$$y_{n+1} = y_n + hf(x_n, y_n) \tag{14.11}$$

In this Exercise, Euler's method is used to find an approximate solution of the differential equation $dy/dx = xy$, with the initial value $y(0) = 1$. The approximation is compared to the analytical solution $y = \exp(x^2/2)$.

◢	A	B	C	D	E	F
1	Euler's Method					
2	Differential equation y'=xy with boundary condition y(0)=1					
3		h=	0.1			
4						
5	i	x	y	h*f(x,y)	exact	
6	0	0.00	1.00000	0.00000	1.00000	
7	1	0.10	1.00000	0.01000	1.00501	
8	2	0.20	1.01000	0.02020	1.02020	
9	3	0.30	1.03020	0.03091	1.04603	
10	4	0.40	1.06111	0.04244	1.08329	
11	5	0.50	1.10355		1.13315	

Figure 14.2

(a) Open a new workbook. Enter the text shown on A1:E5 of Figure 14.2. Enter the values in A6:A11.

(b) In C2 enter 0.1 for the value of h. Name the cell as h.

(c) In B6 enter 0, the initial value of x. In C6 enter 1.0 for the initial value of y from the condition $y(0) = 1$; this corresponds to the first term on the right of Equation 14.11 when $n = 0$.

(d) In D6 enter =h*(B6*C6). This corresponds to the second term on the right in Equation 14.11 when $n = 0$. The parentheses are not essential here but are used to make it clear that we are computing the value *h*function*.

(e) In B7 enter =B6 + h to increment *x*. In C7 we compute the first approximation of *y* with =C6 + D6. This corresponds to the left-hand side of Equation 14.11. Copy D6 down to D7.

(f) Copy the cells B7:D7 down to row 11. This computes the successive *y* approximations. Since we shall not be using the last value of $h*f(x,y)$, delete D11.

(g) So that we can compare our approximations in the C column with the exact solution, in E6 enter =EXP(B6^2 / 2) and copy this down to E11. Save the workbook as Chap14.xlsm.

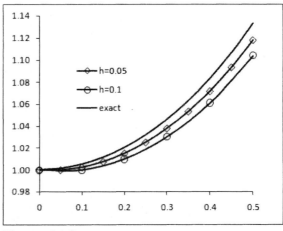

Figure 14.3

Clearly, our answer in the C column is not in very good agreement with the exact values in the E column. A better approximation may be obtained by reducing the size of *h*, the increment for the *x*-values. Figure 14.3 graphs the exact solution, the approximations with $h = 0.1$ and those with $h = 0.05$. This shows that (i) the deviation from the exact values increases with each iteration of Equation 14.11 as expected, and (ii) decreasing the size of *h* significantly improves the solution values.

(h) Modify your worksheet to find the approximations of this differential equation for *x*-values from 0 to 0.5 with steps of 0.025. Save the workbook.

In this example we have used the "crude" Euler method in which the integral in Equation 14.7 was approximated to the area of a rectangle. In the modified Euler method it is approximated to a trapezoid. Compared to the original Euler method, this requires fewer calculations for comparable accuracy. We shall not examine the modified method. The next Exercise uses a more modern method.

Exercise 2: The Runge–Kutta Method

The fourth-order method is sometimes called the *Kutta–Simpson* formula since, when the right-hand side of the differential equation is a function of *x* alone, it reduces to Simpson's ⅓ rule.

In the next chapter, a user-defined function is developed to perform the Runge–Kutta computation.

Like the Euler method, the Runge–Kutta methods finds that an approximation for *y* is based on the previous value. These mathematicians developed a number of algorithms to solve differential equations. We shall use the fourth-order Runge–Kutta method, the derivation of which is beyond the scope of this book. The iterative formula is given in Equation 14.12.

$$y_{n-1} = y_n + \tfrac{1}{6}(k_1 + 2k_2 + 2k_3 + k_4)$$
$$k_1 = hf(x_n, y_n)$$
$$k_2 = hf(x_n + \tfrac{1}{2}h, y_n + \tfrac{1}{2}k_1) \qquad (14.12)$$
$$k_3 = hf(x_n + \tfrac{1}{2}h, y_n + \tfrac{1}{2}k_2)$$
$$k_4 = hf(x_n + h, y_n + k_3)$$

This may look somewhat formidable so let us see how we can put it into a worksheet. We have seen in Exercise 1 how to evaluate the equivalent to k_1; this is the value of the differential function for various *x*- and *y*-values. The second parameter, k_2, is similar except that the *x*-value is incremented by *h* while the *y*-value is incremented by k_1. Each parameter increments *y* by a multiple of the parameter preceding it.

In the previous Exercise we solved *dy/dx = xy* with the initial value *y*(0) = 1 using Euler's method. Here we solve the same problem using the Runge–Kutta method so that we may compare the results.

(a) Open the workbook Chap14.xlsm and make Sheet2 active. Enter the text shown on A1:H5 of Figure 14.4.

(b) Name the cell C4 as h.

(c) Enter the series of values in A6:A11 and B6:B11.

(d) In C6 enter the value 1.0. This is the initial condition *y*(0) = 1 and corresponds to the first term to the right in Equation 14.12 for *n* = 0.

(e) Enter these formulas to compute the *k* parameters:

D6: =h*(B6 * C6)

E6: =h*((B6 + h/2) * (C6 + D6/2))

F6: =h*((B6 + h/2) * (C6 + E6/2))

G6: =h*((B6 + h) * (C6 + F6))

◢	A	B	C	D	E	F	G	H
1	Runge-Kutta Method							
2	Differential equation y' = x*y with boundary condition y(0)=1							
3								
4		h =	0.1					
5	i	x	y	k1	k2	k3	k4	Error
6	0	0.0	1.00000	0.00000	0.00500	0.00501	0.01005	0.000E+00
7	1	0.1	1.00501	0.01005	0.01515	0.01519	0.02040	-2.607E-11
8	2	0.2	1.02020	0.02040	0.02576	0.02583	0.03138	-2.684E-10
9	3	0.3	1.04603	0.03138	0.03716	0.03726	0.04333	-1.023E-09
10	4	0.4	1.08329	0.04333	0.04972	0.04987	0.05666	-2.854E-09
11	5	0.5	1.13315					-6.949E-09

Figure 14.4

We have shown that for the equation $dy/dx = xy$, the Runge–Kutta method is clearly far superior to Euler's method. It may be shown that this is true for all equations.

(f) In C7 enter =C6 + (1/6)*(D6 + 2*E6 + 2*F6 + G6) to compute the first approximation. Compare this formula with Equation 14.12.

(g) Copy the cells D6:G6 to D7:G7. Then copy C7:G7 down to row 10 and copy C10 to C11. This computes the successive *y* approximations.

(h) To compute the error in our approximations enter in H6 =C6 -EXP(B6^2 / 2), format as shown and copy down to H11. Save the workbook Chap14.xlsm.

Exercise 3: Solving with a User-Defined Function

In the previous Exercise we solved $dy/dx = xy$. We would need to make many edits to the worksheet to solve for another equation $dy/dx = f(x,y)$. If we put the function $f(x,y)$ in a module, we need edit only the module (and the initial value) to change our worksheet.

In this Exercise we find the values of *y* that satisfy the equation $dy/dx = 1/(x + y)$ with the initial value $y(0) = 2$. We use *x*-values from 0 to 1.0 in increments of 0.2.

The completed worksheet (Figure 14.5) is very similar to that in the previous exercise. To save time, you may wish to copy that worksheet to a new sheet and edit it to agree with the instructions below. Any of the Copy and Paste methods may be used to duplicate the workbook. Alternatively, hold down Ctrl and drag the tag of the sheet to be copied to the right. When Ctrl is released, a copy of the sheet is inserted into the workbook. Right click the tag and rename it.

◢	A	B	C	D	E	F	G
1	Runge-Kutta Method						
2	Solving dy/dx = f(x,y) with a user-defined function						
3							
4	increment	h	0.2				
5	initial x value	x0	0				
6	initial y value	y0	2				
7	i	x	y	k1	k2	k3	k4
8	0	0.0	2.00000	0.10000	0.09302	0.09317	0.08722
9	1	0.2	2.09327	0.08721	0.08207	0.08216	0.07766
10	2	0.4	2.17549	0.07766	0.07368	0.07374	0.07019
11	3	0.6	2.24927	0.07019	0.06702	0.06705	0.06418
12	4	0.8	2.31636	0.06418	0.06157	0.06159	0.05921
13	5	1	2.37797				

Figure 14.5

(a) Open the VBE and insert a module on the Chap14 project. Enter this function on the module sheet:

```
Function RKfunc(x, y)
      RKfunc = 1 / (x + y)
End Function
```

(b) On Sheet3 enter the text and values shown in A1:G7 of Figure 14.5. Name the cells C4:C6 using the text in B4:B6.

(c) Enter the series of values in A8:A13.
Enter these formulas:
B8: =x0   The initial x value
C8: =y0   The initial y value
D8: =h*rkfunc(B8, C8)   The k parameters
E8: =h*rkfunc ((B8 + h/2), (C8 + D8/2))
F8: =h*rkfunc ((B8 + h/2), (C8 + E8/2))
G8: =h*rkfunc ((B8 + h), (C8 + F8))

(d) In B9 enter =B8+h to increment x.

(e) In C9 enter =C8 + (1/6)*(D8 + 2*E8 + 2*F8 + G8) to compute the first approximation for y.

(f) Copy D8:G8 to line 9. In row 9, we have the second y-value, and the k values needed to compute the third y-value.

(g) Copy B9:G9 down to line 13 to compute the successive y-values. Save the workbook.

If your values do not agree with Figure 14.5 you need to check the function in the module and the formulas on the worksheet. Remember that formulas can be displayed with [Ctrl]+[`]. To check the function, move to a blank cell such as A20 and enter =rkfunc(3,1). This should return the value 0.25.

(h) Now that you have solved one equation, modify the module sheet and the values in the named cells of the worksheet to solve the equation $dy/dx = x^2 + y$ with the initial value $y(1) = 1$. Find the value of y when $x = 1.5$ using first $h = 0.1$, then $h = 0.01$. You will need to extend the worksheet in the second case. The analytical result to eight decimal places is $y(1.5) = 2.64232762$.

Simultaneous and Second-Order Differential Equations

Consider a pair of simultaneous equations having the form:

$$y' = g(x,y,z)$$
$$u' = f(x,y,z) \tag{14.13}$$

The Runge–Kutta formulas for these equations are given in Equation 14.14.

$$y_{n-1} = y_n + \tfrac{1}{6}(k_1 + 2k_2 + 2k_3 + k_4)$$
$$u_{n-1} = u_n + \tfrac{1}{6}(q_1 + 2q_2 + 2q_3 + q_4)$$
$$k_1 = hg(x_n, y_n, u_n)$$
$$q_1 = hf(x_n, y_n, u_n)$$
$$k_2 = hg(x_n + \tfrac{1}{2}h, y_n + \tfrac{1}{2}k_1, u_n + \tfrac{1}{2}q_1)$$
$$q_2 = hf(x_n + \tfrac{1}{2}h, y_n + \tfrac{1}{2}k_1, u_n + \tfrac{1}{2}q_1) \tag{14.14}$$
$$k_3 = hg(x_n + \tfrac{1}{2}h, y_n + \tfrac{1}{2}k_2, u_n + \tfrac{1}{2}q_2)$$
$$q_3 = hf(x_n + \tfrac{1}{2}h, y_n + \tfrac{1}{2}k_2, u_n + \tfrac{1}{2}q_2)$$
$$k_4 = hg(x_n + h, y_n + k_3, u_n + q_3)$$
$$q_4 = hf(x_n + h, y_n + k_3, u_n + q_3)$$

Equations of order greater than one may be solved by transforming them into sets of simultaneous equations. For example, to solve $y'' = ay' + by + c$, we make the substitution $y' = u$. The introduction of the auxiliary variable u allows us to write the second-order equations as two simultaneous equations:

$$y' = u$$
$$u' = au + by + c \tag{14.15}$$

Comparing Equations 14.14 and 14.15, we see that g is a function only of u and is a very simple function: it has the value of u. This simplifies the k terms in Equation 14.14:

$$k_1 = h(u_n)$$
$$k_2 = h(u_n + \tfrac{1}{2}q_1)$$
$$k_3 = h(u_n + \tfrac{1}{2}q_2) \qquad (14.16)$$
$$k_4 = h(u_n + q_3)$$

Exercise 4: Solving a Second-Order Equation

In this Exercise we apply the equations developed above to solve:
$$y'' = y' + y = \sin(x)$$
with boundary conditions $y(0) = 0$ and $y'(0) = 0$

Our task is to obtain approximate values of y and y' when $x = 1$. With the substitution $y' = u$, we get a pair of equations:

$$y' = u \qquad \text{initial value } y(0) = 0$$
$$u' = \sin(x) - y - u \qquad \text{initial value } u(0) = 0$$

Comparing these with Equation 14.13, we see that $g = u$, so we will use the simplified k values of Equation 14.16. We also see that $f = \sin(x) - y - u$.

The function f is referenced in each of the q terms, so it will be more convenient to use a module function. Furthermore, by changing the module you will be able to use the same worksheet for another function.

◢	A	B	C	D	E	F	G	H	I	J	K
1	Second-order differential equation						$y'' + y' + y = \sin(x)$				
2							$y''(0) = y(0) = 0$				
3	xinit	yinit	uinit	h							
4	0	0	0	0.2							
5											
6	x	y	u	k1	q1	k2	q2	k3	q3	k4	q4
7	0.0	0.000	0.000	0.000	0.000	0.000	0.020	0.002	0.018	0.004	0.036
8	0.2	0.001	0.019	0.004	0.036	0.007	0.051	0.009	0.049	0.014	0.062

Figure 14.6

(a) With Chap14.xlsm open, go to the VBE and insert a new module. For this exercise, code the function:

```
Function f(x, y, u)
    f = Sin(x) - y - u
End Function
```

(b) Move to Sheet4 and enter the text and values shown in A1:K6 of Figure 14.6.

(c) Select A3:D4 and name the cells in row 4.

(d) The formulas in row 7 are as follows.
   A7: =xinit
   B7: =yinit
   C7: =uinit
   D7: =h*C7
   E7: =h*f(A7,B7,C7)
   F7: =h*(C7+E7/2)
   G7: =h*f(A7+h/2,B7+D7/2,C7+E7/2)
   H7: =h*(C7+G7/2)
   I7: =h*f(A7+h/2,B7+F7/2,C7+G7/2)
   J7: =h*(C7+I7)
   K7: =h*f(A7+h,B7+H7,C7+I7)

(e) The formulas in row A8:C8 are shown here. Those in columns D to K may be copied from the row above.
   A8: =A7+h
   B8: =B7+(D7+2*F7+2*H7+J7)/6
   C8: =C7+(E7+2*G7+2*I7+K7)/6
   D8: =h*C8

(f) Copy row 8 down to row 12 to get a final value of $x = 1.0$. The results should be $y(1.0) = 0.119394$ and $y'(1.0) = 0.307960$.

(g) Try other values of h such as 0.1 and 0.05 to see if the approximations converge. You will need to expand the table to have $x = 1.0$ in the final row.

Exercise 5: The Simple Pendulum

The equation of motion for a simple pendulum of length L is

$$\frac{d^2\theta}{dt^2} - \frac{g}{L}\sin(\theta) = 0$$

Most textbooks consider a pendulum that starts with a small displacement and use the approximation $\sin(\theta) \approx \theta$. Our approximation will be to use the Runge–Kutta method to solve this second-order differential equation to show how the angle and angular velocity change with time. We will model a 0.75-meter pendulum which is started with a displacement of 0.8 radians from the perpendicular.

As before, we start with the substitution $d\theta/dt = u$, giving:

$$\theta' = u \qquad\qquad \theta\,(t = 0) = 0.8$$
$$u' = -(g\,/\,L)\sin(\theta) \qquad\qquad u\,(t = 0) = 0$$

(a) On the same module used for Exercise 4, code the Pend function

```
Function Pend(L, angle )
    g = 9.8
    Pend = (-g/L) * Sin(angle)
End Function
```

The parentheses around g/L help in reading the formula.

(b) On Sheet5 enter the text and values shown in A1:K6 of Figure 14.7.

◢	A	B	C	D	E	F	G	H	I	J	K
1	Pendulum										
2											
3	InitTime	InitAngle	InitVel	Length	h						
4	0	0.8	0	0.75	0.1						
5											
6	Time	Angle	Velocity	k1	q1	k2	q2	k3	q3	k4	q4
7	0.0	0.800	0.000	0.000	-0.937	-0.047	-0.937	-0.047	-0.916	-0.092	-0.894
8	0.1	0.753	-0.923	-0.092	-0.894	-0.137	-0.849	-0.135	-0.827	-0.175	-0.758

Figure 14.7

(c) The formulas needed in row 7 start with:

```
A7:  =InitTime
B7:  =InitAngle
C7:  =InitVel
D7:  =h*C7
E7:  =h*Pend(Length, B7)
F7:  =h*(C7+E7/2)
G7:  =h*Pend(Length, B7+D7/2)
```

(d) Using what you learned in Exercise 4, complete the formulas in rows 7 and 8. Copy row 8 down to row 37.

(e) Make a chart showing how the angle and the velocity vary with time.

Problems

1. Use Euler's method to solve the differential equation

    $$\frac{dy}{dx} = x + y$$

    Subject to the initial condition $x(0) = 0, y(0) = 0$. Use steps of 0.1 and 0.05 up to $x = 1$. Compare your results with the analytical solution $y = e^x - x - 1$.

2. *Use Euler's method with h = 0.05 to solve the differential equation $y' = -2xy$ with $y(0) = 1$ for $0 \le x \le 2$. The exact solution is $y = \exp(-x^2)$, but be careful how you make the Excel formula; remember, negation has higher priority than exponentiation.

3. Repeat Problem 2 using the Runge–Kutta method. Graphically compare (Figure 14.8) the errors in each method with $h = 0.05$.

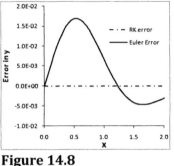

Figure 14.8

4. Use the Runge–Kutta method with $h = 0.1$ to solve the differential equation $y' = -2x - y$ with $y(0) = -1$ for $0 \le x \le 2$.

5. Water flows from an inverted conical tank with a circular orifice at the rate

    $$x'(t) = -0.6\pi r^2 \sqrt{2g}\,\frac{\sqrt{x}}{A(x)}$$

    where r is the radius of the orifice, x is the height of the water level from the vertex of the cone, and $A(x)$ is the area of cross section of the tank x units above the orifice. Find the time when the tank is empty if $r = 0.1$ feet, $g = 32$ feet/sec^2, and the tank was initially filled with $512\pi/3$ cubic feet of water to a level of 8 feet.

6. *The circuit[1] shown in the following figure contains a battery (E), an inductance (L), and a resistor (R) whose magnitude varies with its temperature and hence with the current passing through it. Its resistance can be expressed by $R = a +$

[1] M. L. James et al., *Applied Numerical Methods for Digital Computation*, Harper & Row, New York, 1977 (page 406).

bi, where *a* and *b* are constants and *i* is the current. The switch (S) is closed at time *t* = 0 and the resulting current can be described by the differential equation:

$$\frac{di}{dt} = \frac{E}{L} - \frac{b}{L}i^3 - \frac{a}{L}i$$

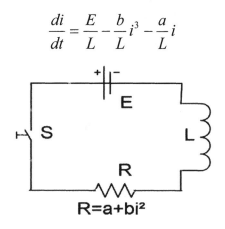

R=a+bi²

Using Exercise 3 as a model, compute the current from *t* = 0 to *t* = 0.8 seconds in increments of 0.001 for the case *E* = 200 volts, *L* = 3 henrys, *a* = 100 ohms, and *b* = 50 ohms/amp². Since the independent variable *t* does not appear explicitly in the differential, the terms for the Runge–Kutta *k*'s will involve only the current *i*.

[2]D. G. Zill, *Differential Equations with Computer Lab Experiments*, PWS Publishing Co., Boston, 1995 (page 86).

7. *If air resistance is proportional to the square of the instantaneous velocity *v*, then the velocity of a mass *m* dropped from a height *h* is determined by[2]

$$m\frac{dv}{dt} = mg - kv^2$$

Use the Runge–Kutta method with increments of 0.5 to find an approximate velocity after 5 seconds for a mass of 5 slugs. Let *g* = 32 and *k* = 0.125. Compare your result with the analytical solution:

$$v(t) = \sqrt{\frac{mg}{k}} \tanh \sqrt{\frac{kg}{m}}t$$

[3]Hint: Imagine the outflow can be paused. Let 10 gals flow in; what is the new concentration? Let the 10 gals escape in a flash. How much salt remains? Do this for successive minutes. Can a single formula be used to find the amount after *n* minutes? Repeat the solution with smaller time intervals.

8. A 500-gal tank is filled with water with 20 lbs of dissolved salt. Fresh water flows in at 10 gal/min. How long will it take until there are just 5 lbs of salt in the tank? Assume perfect mixing. Solve[3] this without writing down a differential equation; see Figure 14.9. Can you write a formula to compute *t* when *Salt* = 5? As Δt gets smaller, your answer will converge to a more accurate value.

◢	A	B	C	D	E	F	G
1	Tank Problem						
2							
3	Volume	500					
4	S_start	20					
5	S_end	5					
6	Flow	10					
7	dt	0.5					
8							
9	n	t	salt		n	t	salt
10	0	0	20		0	0	20
11	1	0.5	19.802		20	10	16.391
12	2	1	19.606		40	20	13.433
13	3	1.5	19.412		60	30	11.009
14	4	2	19.220		80	40	9.022
15	5	2.5	19.029		100	50	7.394
16	6	3	18.841		120	60	6.060
17	7	3.5	18.654		140	70	4.966

Figure 14.9

9. *Write a differential equation for Problem 8. Integrate this analytically. Make a plot of *Salt* against *time* from the data in Problem 8 and compare its trendline to that predicted by your analytical solution.

10. Suppose a ship moving at speed 6 m/s suffers a sudden loss of power. We will assume the distance s (meters) it moves in time t (seconds) is governed by the differential equation[4]

$$\frac{ds}{dt} = v_0 \exp\left(-\frac{kt}{m}\right)$$

with initial condition $s(0) = 0$. Use the Runge–Kutta method to find how far it will move in the first 60 seconds if $k = 44 \times 10^3$ kg/s and $m = 2.55 \times 10^6$ kg. For your first attempt use steps of $h = 10$ seconds. Then repeat using steps of $h = 5$ and $h = 1$. Compare your approximations with the exact values computed with $s(t) = \frac{v_0 m}{k}\left(1 - \exp(-\frac{kt}{m})\right)$.

[4] J. R. Hanly, *Essential C++ for Engineers and Scientists*, Addison-Wesley, Reading, MA, 1977 (page 362).

11. Getting your information from a textbook and/or the Internet, repeat Problem 4 using (i) integration using the Taylor series method and (ii) the Runge–Kutta–Fehlberg method. The second method is used in programs such as MathLab and Maple for their OED routines.

15

Modeling II

This chapter will give us the opportunity to model some practical problems using what we have learned in the last three chapters.

Exercise 1: The Four-Bar Crank

In this Exercise we examine an engineering mechanism used to generate a complex rotational motion from a simple one motion.

The four-bar mechanism (see Figure 15.1) consists of three movable links (a, b, and c) and a fixed link d. The link a is rotated, causing link c to rotate. Our objective is to map the relationship between the angles θ and φ.

There is an interesting dynamic demonstration of the four-bar crank at: http://www.dim.unipd.it/lot/mbsymba/kinematics/kin_4bar_linkage.html

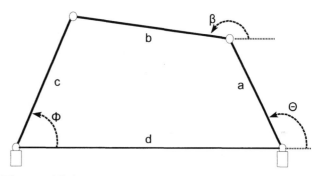

Figure 15.1

For the quadrilateral formed by the four links, the algebraic sum of the vertical component and the algebraic sum of the horizontal component must equate to zero. This gives the two equations:

$$a \sin\theta + b \sin\beta - c \sin\phi = 0$$
$$a \cos\theta + b \cos\beta - c \cos\phi + d = 0$$

Adding the squares of these gives the Freudenstein equation:

$$R_1 - R_2 \cos\phi + R_3 - \cos(\theta - \phi) = 0$$

where:

$$R_1 = d/c$$
$$R_2 = d/a$$
$$R_3 = (a^2 - b^{-2} + c^2 + d^2)/2ac$$

We will use Microsoft Excel's Solver to find the output angle for input angles in the range 0 to 360° in 5° steps.

◢	A	B	C	D	E	F	G	H
1	Four-bar Crank			Input Angle		Output Angle		Freudenstein's
2				Degrees	Radian	Degrees	Radian	Equation
3				0	0	57.3	1	0.6291
4	Crank lengths			5	0.0873	57.3	1	0.5540
5	a	1		10	0.1745	57.3	1	0.4760
6	b	2		15	0.2618	57.3	1	0.3956
7	c_	2		20	0.3491	57.3	1	0.3136
8	d	2		25	0.4363	57.3	1	0.2304
9	Ratio1	1		30	0.5236	57.3	1	0.1468
10	Ratio2	2		35	0.6109	57.3	1	0.0633
11	Ratio3	1.25		40	0.6981	57.3	1	-0.0193
12				45	0.7854	57.3	1	-0.1006
13	Unsolved			50	0.8727	57.3	1	-0.1797

Figure 15.2

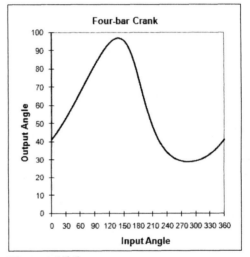

Figure 15.3

(a) Start a new workbook and on Sheet1 enter all the text shown in Figure 15.2. Cell A13 contains a formula, not text.

(b) Enter the specifications for the four-bar crank in B5:B8.

(c) Select A5:B8 and name the cells B5:B8.

(d) Compute the *ratio* values with:
   B9: =d/c_
   B10: =d/a
   B11: =(a^2 -b^2 + c_^2 + d^2)/(2*a*c_)

(e) Select A9:B11 and name the cells in B9:B11.

(f) In D3 enter the value 0 and in D4 enter 5. Select these two cells and by dragging the fill handle down to D75 make the series 0 to 360 in increments of 5.

(g) In E3 enter =RADIANS(D3) and double click the fill handle to fill the formula down to F75.

(h) In F3 enter the formula =DEGREES(G3) and in G3 enter the value 1. Select both cells and double click the fill handle to fill down to row 75.

(i) In H3 enter: =Ratio1*COS(E3) - Ratio2*COS(G3) + Ratio3 - COS(E3-G3). Refer to the Freudenstein equation above to ensure you have this correct.

(j) Give G3:G75 the name *Output* and H3:H75 the name *Equation*.

(k) In A13 enter: =IF(SUM(Equation)>0.00001, "Unsolved", "Solved"). Merge and Center this across A13:B13. This formula will display *Unsolved* until we invoke Solver and solve all 73 Freudenstein equations.

(l) Save the workbook as Chap15.xlsm as a precaution. Make it macro-enabled since we shall be adding modules later.

We are ready to have Solver make every cell in the *Equation* range equal to zero by changing the *Output* values.

(m) Call up Solver with *Data | Analysis | Solver*. Clear the *Target* box; in the *By Changing Cells* box enter Output and add the *Constraint* Equation=0. Click the Solve button.

(n) After about five seconds (watch the Excel status bar as it displays messages like *Trial Solution 42*), Solver will have completed its task.

(o) Make a plot to show how the two angles are related as in Figure 15.3 Save the workbook.

The companion website has a workbook called *FourBarCrank.xlsm*, which contains VBA code to make an animated diagram using the results of this exercise.

Exercise 2:
Temperature Profile
Using Matrix Algebra

Consider a thin metal sheet (Figure 15.4) whose edges are maintained at specified temperatures and the sheet is allowed to come to thermal equilibrium. Our task is to compute the approximate temperatures at various positions on the plate.

We need to make some assumptions. The first is that the two faces of the plate are thermally insulated. Thus there is no heat transfer perpendicular to the plate. The second assumption starts with the mean-value theory, which states: if P is a point on a plate at thermal equilibrium and C is a circle centered on P and completely on the plate, then the temperature at P is the average value of the temperature on the circle. The calculations required to use this theory are formidable, so we will use an approximation. We shall consider a finite number of equidistant points on the plate and use the discrete mean-value theory, which states that the temperature at point P is the average of the temperatures of P's nearest neighbors.

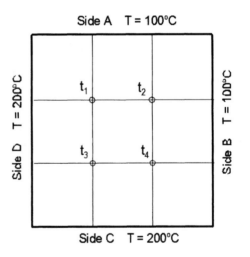

Side A   T = 100°C

Side D   T = 200°C

Side B   T = 100°C

t_1 t_2

t_3 t_4

Side C   T = 200°C

Figure 15.4

The most convenient way to arrive at the equidistant point is to divide the plate using equally spaced vertical and horizontal lines. In Figure 15.4 two such lines have been drawn parallel to each axis. This gives four interior points for the calculation. With such a small number, the results will not be very accurate. However, the methodology is the same regardless of the number of points, and it is simpler to describe and test the method initially with four points.

Applying the averaging rule, the temperatures of the four interior points are given by:

$$t_1 = (100 + t_2 + t_3 + 200)/4$$
$$t_2 = (100 + 100 + t_4 + t_1)/4$$
$$t_3 = (t_1 + t_4 + 200 + 200)/4 \qquad (15.1)$$
$$t_4 = (t_2 + 100 + 200 + t_3)/4$$

Equation 15.1 may be written in a more general form using variables $a, b, c,$ and d rather than numerical values. It may then be rearranged in the form:

$$t_1 = (t_2 + t_3)/4 + (a + d)/4$$
$$t_2 = (t_4 + t_1)/4 + (a + b)/4$$
$$t_3 = (t_1 + t_4)/4 + (c + d)/4 \qquad (15.2)$$
$$t_4 = (t_2 + t_3)/4 + (b + c)/4$$

To be able to use a matrix method, each equation in Equation 15.2 must have the same form:

$$t_1 = (0.00t_1 + 0.25t_2 + 0.25t_3 + 0.00t_4) + (a + d)/4$$
$$t_2 = (0.25t_1 + 0.00t_2 + 0.00t_3 + 0.25t_4) + (a + b)/4$$
$$t_3 = (0.25t_1 + 0.00t_2 + 0.00t_3 + 0.25t_4) + (c + d)/4 \qquad (15.3)$$
$$t_4 = (0.00t_1 + 0.25t_2 + 0.25t_3 + 0.00t_4) + (b + c)/4$$

We may write this system of four equations as:

$$T = MT + B \qquad (15.4)$$

where:

$$T = \begin{bmatrix} t_1 \\ t_2 \\ t_3 \\ t_4 \end{bmatrix}, \quad M = \begin{bmatrix} 0 & 0.25 & 0.25 & 0 \\ 0.25 & 0 & 0 & 0.25 \\ 0.25 & 0 & 0 & 0.25 \\ 0 & 0.25 & 0.25 & 0 \end{bmatrix}, \quad B = \begin{bmatrix} (a+d)/4 \\ (a+b)/4 \\ (c+d)/4 \\ (b+c)/4 \end{bmatrix}$$

We can rearrange Equation 15.4 in this way

$$T - MT = B$$
$$(I - M)T = B$$

where I is the identity matrix, a matrix in which the diagonal elements are 1 and the off-diagonal elements are 0.

So to solve for T we use:

$$T = (I - M)^{-1} B \qquad (15.5)$$

(a) On Sheet2 of Chap15.xlsm, enter all the text shown (Figure 15.5).

(b) Enter the temperature values in A4:D4. Select A3:D4 and name the cells A4:D4. Enter values as shown in A7:D17.

(c) In A20 enter the formula =A14 - A8. Copy this to A20:D23 to compute the elements of [*I* – *M*].

◢	A	B	C	D	E	F	G	H	I
1	Temperature Profile			Using Matrix Method					
2									
3	SideA	SideB	SideC	SideD					
4	100	100	200	200					
5									
6									
7		M matrix					Inverse of I - M		
8	0	0.25	0.25	0		1.167	0.333	0.333	0.167
9	0.25	0	0	0.25		0.333	1.167	0.167	0.333
10	0.25	0	0	0.25		0.333	0.167	1.167	0.333
11	0	0.25	0.25	0		0.167	0.333	0.333	1.167
12									
13		I matrix				B matrix		T matix	
14	1	0	0	0		75		T1	150
15	0	1	0	0		50		T2	125
16	0	0	1	0		100		T3	175
17	0	0	0	1		75		T4	150
18									
19		I - M							
20	1	-0.25	-0.25	0					
21	-0.25	1	0	-0.25					
22	-0.25	0	1	-0.25					
23	0	-0.25	-0.25	1					

Figure 15.5

(d) Select F8:I11 and type the formula =MINVERS(A20:D23). Press ⌈Ctrl⌋+⌈⇧ Shift⌋+⌈Enter⌋ to complete the array formula. This computes [*I* – *M*]$^{-1}$.

(e) The formulas for the *B* matrix are:
F14: =(SideA + SideD)/4
F15: =(SideA + SideB)/4
F16: =(SideC + SideD)/4
F17: =(SideB + SideC)/4

(f) All that remains is to multiply [*I* – *M*]$^{-1}$ by *B*. With H14:I17 selected, enter =MMULT(F8:I11, F14:F17) and commit the array formula with ⌈Ctrl⌋+⌈⇧ Shift⌋+⌈Enter⌋.

(g) Save the workbook.

Exercise 3: Temperature Profile using Solver

In this Exercise we solve the same problem as in Exercise 2, but here we shall use Solver and a 5 × 5 mesh rather than 2 × 2. Looking at the before (Figure 15.6) and after (Figure 15.7) screen captures of the worksheet will make it clearer for the reader what has to be done.

The *Model* range (C4:G8) and its borders are numeric values. The *Solution* range (C12:G16) has formulas calculating the average of each cell's four neighbors; for example, C12 has the formula =AVERAGE(C3,B4,D4,C5). We will have Solver change each of the *Model* cells until they equal their corresponding *Solution* cell. The trick is that the *Solution* range contains formulas. This may sound a little like a circular reference, but it really is not.

◢	A	B	C	D	E	F	G	H
1	Temperature profile of metal plate					Unsolved		
2								
3		Model	100	100	100	100	100	
4		200	100.0	100.0	100.0	100.0	100.0	100
5		200	100.0	100.0	100.0	100.0	100.0	100
6		200	100.0	100.0	100.0	100.0	100.0	100
7		200	100.0	100.0	100.0	100.0	100.0	100
8		200	100.0	100.0	100.0	100.0	100.0	100
9			200	200	200	200	200	
10								
11		Solution	100	100	100	100	100	
12		200	125.0	100.0	100.0	100.0	100.0	100
13		200	125.0	100.0	100.0	100.0	100.0	100
14		200	125.0	100.0	100.0	100.0	100.0	100
15		200	125.0	100.0	100.0	100.0	100.0	100
16		200	150.0	125.0	125.0	125.0	125.0	100
17			200	200	200	200	200	

Figure 15.6

(a) On Sheet3 of Chap15.xlsm copy all the text and numbers shown in rows 1 through 9 in Figure 15.6 with the exception of F1.

(b) Select C4:G8 and in the Name box enter the word Model so as to name that range.

(c) In F1 enter =IF(C4=C12,"Solved","Unsolved"). This is our "flag" to tell us if Solver needs to be called should we alter the model.

(d) We want the border of the *Solution* range to match those of the *Model* range. In C11 enter =C3 and copy across to G3. Do the same for the other three borders.

(e) In C12 enter the formula =AVERAGE(C3,B4,D4,C5) and fill this down and across to G16.

(f) Select C12:G16 and give it the name Solution. Enter the same text in C11.

(g) Now we call Solver. Clear the *Target* cell; in the *By Changing Cells* enter Model; and add the *Constraint* Solution=Model. Click the *Solve* button to generate the results shown in Figure 15.7.

(h) Save the workbook.

⬛	A	B	C	D	E	F	G	H
1	Temperature profile of metal plate					Solved		
2								
3	Model		100	100	100	100	100	
4		200	150.0	129.9	119.6	112.5	106.3	100
5		200	170.1	150.0	136.0	124.2	112.5	100
6		200	180.4	164.0	150.0	136.0	119.6	100
7		200	187.5	175.8	164.0	150.0	129.9	100
8		200	193.7	187.5	180.4	170.1	150.0	100
9			200	200	200	200	200	
10								
11	Solution		100	100	100	100	100	
12		200	150.0	129.9	119.6	112.5	106.3	100
13		200	170.1	150.0	136.0	124.2	112.5	100
14		200	180.4	164.0	150.0	136.0	119.6	100
15		200	187.5	175.8	164.0	150.0	129.9	100
16		200	193.7	187.5	180.4	170.1	150.0	100
17			200	200	200	200	200	

Figure 15.7

You can experiment by changing the temperature setting for the four borders of the plate and rerunning Solver.

In earlier editions of this book a circular reference method was used, but it has been abandoned in favor of the more straightforward Solver technique. The interested reader will find details of the circular reference method on the companion website in *CircularReference.xlsm*.

Exercise 4: Emptying the Tank

In this Exercise we solve a simple differential equation using the Runge–Kutta method. In Chapter 14 we placed the terms needed for the Runge–Kutta approximation on the worksheet. In this Exercise we use a user-defined function. In the subsequent Exercise a function is used to iterate the approximation.

A cylindrical tank of diameter D has a short pipe of diameter d at the bottom. The tank is initially filled with water to a height h. We wish to examine how changing the diameter of the pipe alters the rate of discharge of the tank. The problem chosen has an analytical solution. You may wish to find it and compare the results from it with those found using the Runge–Kutta approximation.

For a short pipe, we may assume the rate of change of h is:

$$\frac{dh}{dt} = -\frac{d^2}{D^2}\sqrt{2gh}$$

Working in metric units, we shall use $g = 9.8$ ms^{-2}.

We begin by developing a user-defined function to compute dh/dt for any value of h. We would like a worksheet that lets us vary both the diameter of the pipe d and of the tank D. Clearly our equation could be rewritten as $dh/dt = -R^2\sqrt{2gh}$ where $R = d/D$, the ratio of the two diameters. The required function is given in Figure 15.8. This uses the VBA square root function Sqr, not the Excel function SQRT.

In the interests of clarity, dimension statements have been omitted in all our VBA functions. The reader is strongly advised to set the VBEditor to require these. All numeric variable should be dimensioned as *Double*.

```
'The function of the differential equation
Function tank(height, ratio)
    Const g = 9.8
    tank = -(ratio ^ 2) * Sqr(2 * g * height)
End Function
```

Figure 15.8

The VBA function to perform the Runge–Kutta approximation is shown in Figure 15.9. Compare the k expressions with those in Equation 14.12. Since t does not appear to the right in the differential equation we are solving, there are no x terms in our k expressions. The y term of Equation 14.12 becomes the *height* term. What was called h in Equation 14.12, we call *incr* (short for *increment*) in our function. Since *height* is water height value, a variable called h might be confusing. The *ratio* term has been added so that it may be passed to the *tank* function. We must be careful in the worksheet to call this function with the *height, incr,* and *ratio* arguments in the correct order.

The inner parentheses around 2*k1 and 2*k2 in the last statement are not required. However, because of the way VBA inserts spaces, they were added to make the equation more readable.

```
'Function to compute Runge-Kutta approximation
Function RKapprox(height, incr, ratio)
    k1 = incr * tank(height, ratio)
    k2 = incr * tank(height + k1 / 2, ratio)
    k3 = incr * tank(height + k2 / 2, ratio)
    k4 = incr * tank(height + k3, ratio)
    RKapprox = height + (k1 + (2 * k2) + (2 * k3) + k4) / 6
End Function
```

Figure 15.9

◢	A	B	C	D	E
1	Tank Problem		Version 1		
2					
3	Initial height	1.00	meters		
4	Time increment	0.1	secs		
5	Tank diameter	1.00	meters		
6	Pipe diameter	5.00	cm		
7	Ratio of diameters	0.05			
8					
9	Time	Height		Summary	
10	0	1.000		time	height
11	0.1	0.999		0	1.00
12	0.2	0.998		5	0.95
13	0.3	0.997		10	0.89
14	0.4	0.996		15	0.84
15	0.5	0.994		20	0.79
16	0.6	0.993		25	0.74
17	0.7	0.992		30	0.70
18	0.8	0.991		35	0.65
19	0.9	0.990		40	0.61
20	1	0.989		45	0.56

Figure 15.10

(a) Open Chap15.xlsm and invoke the VBE. Insert a module on which to code the two functions shown in Figures 15.8 and 15.9.

(b) On Sheet4, enter the text values shown in columns A through C of Figure 15.10.

(c) Enter the values in B3:B6 and in B7 enter =*C6*0.01/C5*. The 0.01 converts the pipe diameter to meters.

(d) Enter 0 in A10 and in A11 enter =*A10+B4*. Copy this down to row 110. This gives us the time steps.

(e) In B10 enter =*B3* to set the initial height.

(f) In B11 type the formula =*RKapprox(B10,$B$4,$B$7)*. Hopefully, your worksheet returns the value 0.999 in C11. An error value of #NAME! means that the name of the function in the cell does not match that in the module. If B11 shows #VALUE!, check (i) that the arguments in the formula point to the correct value and (ii) that the RKapprox function is correctly coded.

(g) Copy B11 down to row 1110. This is when the trick of double clicking the fill handle comes into its own.

Our data extends over more than 1000 rows and is too much to absorb. We need to make a summary.

(h) Enter the numbers 0 and 5 in D11 and D12, respectively. Select these two values and drag the fill handle down to row 33 to give a last value of 110.

(i) In E11 enter =*VLOOKUP(D11,$A$10:$B$1110,2,TRUE)* and copy this down to E33.

(j) Make a chart from either the Runge–Kutta data or the summary using a smooth XY chart with a line and no markers. Save the workbook.

Exercise 5: An Improved Tank Emptying Model

The worksheet in Exercise 4 has some faults: (i) it uses a great deal of space, and we had to make a summary table, and (ii) at any one time we can see data for only one pipe size. The faults are addressed in this Exercise.

In the last worksheet (Figure 15.10), the formula in B11 was =RKapprox(B10,$B$4,$B$7). In B12 we compute a newer height from the value in B11. This gets repeated. Why store every value on a worksheet? With VBA we could place data on the worksheet after so many iterations. A first attempt is shown in Figure 15.11.

```
' Tentative Function to make n calculations with RKapprox
Function NewHeight(OldHeight, incr, ratio, n)
    For j = 1 to n
        NewHeight = RKapprox(OldHeight, incr, ratio)
        OldHeight = NewHeight
    Next j
End Function
```

Figure 15.11

However, the Runge–Kutta method is not accurate when *OldHeight* gets small compared to *incr* and may result in negative values that are meaningless in this model. We therefore modify the iteration function so that it stops before completing the *n* iterations if *OldHeight* is small. Plan B, the improved function, is shown in Figure 15.12.

```
'Function to make n calculations with RKapprox
Function NewHeight(OldHeight, incr, ratio, n)
  For j = 1 to n
    If OldHeight < 0.0001 Then
        NewHeight = 0
        Exit For
    Else
        NewHeight = RKapprox(OldHeight, incr, ratio)
        OldHeight = NewHeight
    End If
  Next j
End Function
```

Figure 15.12

(a) Open the VBE and on the existing module for Chap15.xlsm code the function shown in Figure 15.12.

(b) Go to Sheet5 of the workbook and enter the text shown in Figure 15.13.

(c) Enter the values in D3:D5 and name the cells with the text to the left.

(d) Enter the values in rows 7 through 9.

◢	A	B	C	D	E	F
1	Tank emptying problem			Version 2		
2						
3	Initial height (m)	h0	1			
4	Time increment	incr	0.1			
5	Iterations	iter	50			
6						
7	Tank diameter (m)		1	1	1	1
8	Pipe diameter (cm)		1	2	5	10
9	Ratio of diameters		0.01	0.02	0.05	0.1
10						
11		Time (sec)	Height (m)			
12		0	1.000	1.000	1.000	1.000
13		5	0.998	0.991	0.945	0.791
14		10	0.996	0.982	0.892	0.606
15		15	0.993	0.974	0.841	0.446
16		20	0.991	0.965	0.791	0.311
17		25	0.989	0.956	0.742	0.199
18		30	0.987	0.948	0.696	0.113
19		35	0.985	0.939	0.650	0.051
20		40	0.982	0.930	0.606	0.013
21		45	0.980	0.922	0.564	0.000

Figure 15.13

(e) In B12 enter the value 0. In B13 enter =B12+(incr*iter) and copy this down to row 32.

(f) In C12 enter =h0 and copy across to D12. In C13 enter =NewHeight(C12,incr, C$9,iter). Copy this across to column F and down to row 21.

(g) Make a chart similar to that in Figure 15.14. Save the workbook.

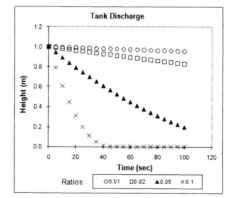

Figure 15.14

Problems

1. In Exercises 4 and 5 we assumed the increment of 0.1 was sufficiently small for accurate results. Copy Sheet5 of Chap15.xlsm by dragging its tab to the right and modify the copied worksheet to use the same height but different *inc* values. To make the comparison easier, each column should use values of *iter* such that *inc* × *iter* = 5. You may wish to use a modification of *NewHeight* to achieve this.

2. As an extension to Exercise 5, write a UDF that estimates the time for the tank to empty. It would be unwise to have the function loop until the height was zero (why?), so we will define "empty" as meaning the height is reduced by at least 99.95%.

3. Use Solver to find the molar volume in Problem 1 in Chapter 11.

4. Use Solver to find the forces in Problem 2 of Chapter 11.

5. Use Solver to find the nodal voltages in Problem 3 in Chapter 11.

6. The accompanying figure shows a network of water pipes.[1] Your task is to develop a Solver model to find the flow of water (Q_i) in each pipe. Two conditions must be satisfied: (i) the algebraic sum of the pressure drops around each closed loop is zero (use the UDF developed in Problem 2 of Chapter 9), and (ii) at any junction, the inflow equals the outflow.

[1]B. Maxfield, *Engineering with Mathcad*, Butterworth-Heinemann, Oxford, UK, 2006 (page 333).

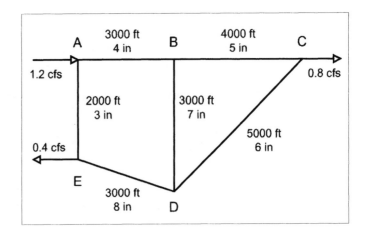

Statistics for Experimenters

In the past there have been some criticisms of the algorithms used by Excel for its statistical functions. Significant improvements were made in Excel 2003 and Excel 2007.

The word *sample* is used in this chapter in the statistical sense. To avoid confusion, we will use *specimen* for the object that is being measured.

Microsoft Excel can be a powerful tool for statistical analysis. In this chapter we look at a very small subset of these tools. The main focus is on the treatment of variability associated with data measurements. Some of the functions introduced are:

AVERAGE — Calculates the arithmetic mean of the values in a data set.

DEVSQ — Calculates the sum of the squares of the deviations of the values from their mean.

FREQUENCY — Calculates how often values in a data set occur within a range of values in a bin.

STDEV — Calculates the sample standard deviation of the values in a data set.

TDIST — Calculates the probability for Student's t–distribution.

TINV — Calculates the t-value of Student's t–distribution.

TTEST — Calculates the probability associated with Student's t–test.

We shall also introduce some of the Data Analysis tools from the Analysis Toolpak.

Exercise 1: Descriptive Statistics

If there is no *Analysis* group tool on the Data tab or no *Data Analysis* tool in the Analysis group, you need to search Help with the word *Toolpak* (spell it just like that) for instructions on how to load the Analysis Toolpak.

An experimenter has collected 100 measurements and wishes to know some statistics of the data set, for example, the average or the sum. To save the task of entering 100 numbers we will have Excel generate some random numbers. The RAND or RANDBETWEEN functions are not appropriate here since they generate uniform distributions of values. So we will use the Random Number generator tool found in the Data Analysis toolbox. To simulate the results from an experiment, we will request random numbers having a normal distribution with a mean (average) of 10 and a standard deviation of 0.5.

(a) Open a new workbook. In A1 of Sheet1 enter the label data. Use the command *Data | Analysis | Data Analysis* and select the item *Random Number Generation*. Complete the dialog box as shown in Figure 16.1. Give A2:A101 the name *data*.

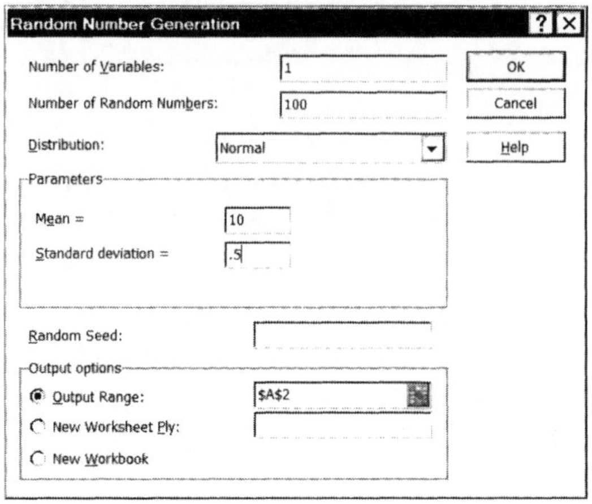

Figure 16.1

The *Input Range* box in Figure 16.2 was filled in by clicking the A column header to give $A:$A which means "all the cells in column A". Since the column contains only our data set this is acceptable and is quicker than selecting A1:A100.

Like all Analysis Toolpak features, Descriptive Statistics generates *values* not *formulas*. This means that its results are static and do not change when the source data is changed.

(b) To generate the statistics quickly, we will use another Data Analysis tool, namely, *Descriptive Statistics*. Complete the dialog box as shown in Figure 16.2. Your worksheet will resemble columns A to D of Figure 16.3 but with different values since we are working with random numbers.

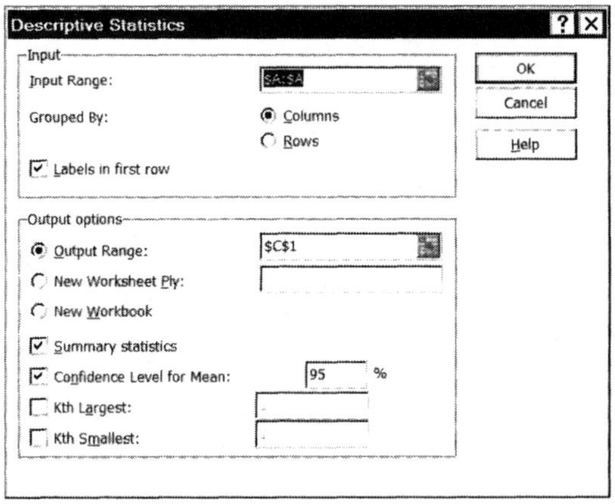

Figure 16.2

(c) There is a worksheet function corresponding to all but two of the statistics generated by the Data Analysis tool. The functions are shown in column F of Figure 16.3. The Confidence Level value returned by the tool differs from that returned by the CONFIDENCE function. We return to this later. Save the workbook as Chap16.xlsx.

◢	A	B	C	D	E	F	G
1	data		data			formulas	
2	10.20012						
3	8.954345		Mean	10.03594	10.03594	=AVERAGE(data)	
4	10.05579		Standard Error	0.05594	0.05594	=STDEV(A2:A101)/SQRT(COUNT(A2:A101))	
5	9.593495		Median	9.99994	9.99994	=MEDIAN(data)	
6	10.31294		Mode	#N/A	#N/A	=MODE(data)	
7	9.838037		Standard Deviation	0.55939	0.55939	=STDEV(data)	
8	11.4453		Sample Variance	0.31291	0.31291	=VAR(data)	
9	9.899985		Kurtosis	-0.08594	-0.08594	=KURT(data)	
10	11.28357		Skewness	0.24594	0.24594	=SKEW(data)	
11	9.116269		Range	2.55930	2.55930	=MAX(data)-MIN(data)	
12	9.633214		Minimum	8.88600	8.88600	=MIN(data)	
13	9.755323		Maximum	11.44530	11.44530	=MAX(data)	
14	10.25377		Sum	1003.59432	1003.59432	=SUM(data)	
15	9.831372		Count	100	100	=COUNT(data)	
16	9.76027		Confidence Level(95.0%)	0.11099	0.00351	=CONFIDENCE(0.95,E7,E15)	

Figure 16.3

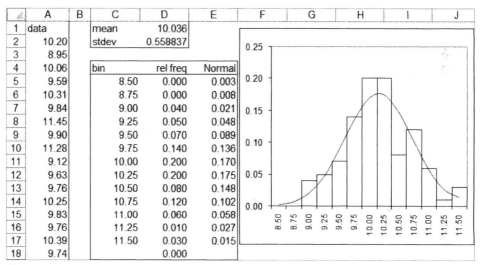

◢	A	B	C	D	E	F	G	H	I	J
1	data		mean	10.036						
2	10.20		stdev	0.558837						
3	8.95									
4	10.06		bin	rel freq	Normal					
5	9.59		8.50	0.000	0.003					
6	10.31		8.75	0.000	0.008					
7	9.84		9.00	0.040	0.021					
8	11.45		9.25	0.050	0.048					
9	9.90		9.50	0.070	0.089					
10	11.28		9.75	0.140	0.136					
11	9.12		10.00	0.200	0.170					
12	9.63		10.25	0.200	0.175					
13	9.76		10.50	0.080	0.148					
14	10.25		10.75	0.120	0.102					
15	9.83		11.00	0.060	0.058					
16	9.76		11.25	0.010	0.027					
17	10.39		11.50	0.030	0.015					
18	9.74				0.000					

Figure 16.4

Exercise 2: Frequency Distribution

Frequently, an experimenter wishes to compare the distribution of experimental data with the normal Gaussian distribution. We will use the data generated in the previous exercise, rounded to two decimal places.

(a) On Sheet2 of Chap16.xlsx, enter the text values shown in Figure 16.4. In A2 enter =ROUND(Sheet1!A2,2). The pointing method works well here. Copy the formula down to row 101. Name the range A2:A101 as data.

(b) Compute the mean and the standard deviations of the data using the formulas =AVERAGE(data) and =STDEV(data) in D1 and D2, respectively. Give these cells the names mean and stdev, respectively.

(c) Enter the values in C5:C17. These will serve as the *bin* values or value intervals for the frequency formula.

Recall that we often use one additional cell for the FREQUENCY function in case there are values greater than our highest bin value.

(d) With D5:D18 selected, enter =FREQUENCY(data, C5:C17)/COUNT(data) and use ⎡⇧ Shift⎤ + ⎡Ctrl⎤ + ⎡Enter ←⎤ to complete the formula. This gives us the relative frequency of each bin value in the data set. Remember, we are using random numbers, and so your data will not be exactly the same as in the figure.

(e) To compute the expected frequencies use these formulas:
E5: =NORMDIST(C5, mean, stdev, TRUE)
E6: =(NORMDIST(C6,mean,stdev,TRUE)
   - NORMDIST(C5, mean, stdev, TRUE))
Copy this down to row 16, and enter the last formula
E17: =(1 - NORMDIST(C16, mean, stdev, TRUE))

(f) You may wish to use SUM(D5:D18) to check that the sum of the relative frequencies are in good agreement with the sum of the expectation values =SUM(E5:E17).

In E5 we compute the cumulative probability for an *x*-value of 8.5 or less. In E6 to E16 we compute the probability for a range of values given by two consecutive *x*-values. In E17 we find the expected probability for values of 11.5 and over.

(g) We are about to make a Line chart of the data in C5:E17. Excel will mistakenly think column C contains a data series rather than being the *x*-category values. To overcome this, temporarily delete the label in C4. Select C4:D17 and make a Line chart using the first subtype.

(h) Right click on the relative frequency data series. Use the *Change Series Chart Type* command to set this to a column type.

(i) Right click this column data series; use the Format Data Series command and, on the Series Options tab, set the *Gap Width* to 0 (*No Gap*).

(j) Format the other data series as a smooth line on the *Line Style* tab.

(k) Save the workbook.

In Problem 3 at the end of the chapter, another way is shown to fit the histogram to the normal curve. It uses somewhat simpler formulas but gives a less accurate result.

Exercise 3: Confidence Limits

Throughout this chapter we concentrate on a two-tailed test. This is appropriate when we compare two means to see whether or not they are unequal. A one-tailed test is appropriate when we need more than this and wish to know, for example, if one mean is really greater than another. A statistic text should be consulted for more information.

In the days before computers (BC), *t*-values had to be found from printed tables. If you compare the results from TINV with a table you should note that TINV gives the value of the two-tailed test statistic.

Measurements are often repeated a number of times, and the results can be summarized by a single value, average, which statisticians call the *mean*. In addition to reporting the mean value, it is often necessary to indicate the spread of the measurements. The most commonly used measure of spread in a data set is the *standard deviation*. Statisticians speak of *population* and *sample* standard deviations. However, in theory, a measurement could be repeated an infinite number of times. Our actual measurements are a subset (a *sample*) of this hypothetical set, so we use the sample standard deviation. Hence the appropriate Microsoft Excel function is STDEV rather than STDEVP.

The data in column A of Figure 16.4 might be reported as *the value of x was found to be 10.036 ± 0.559 (n = 100)*. In many cases, it would be appropriate to use only two decimal places since that was the precision of the raw data.

Our multiple measurements of specimens could be used to estimate a mean value ($\bar{x}$), while our goal is to determine the population (true) average is μ. We would like some way of expressing how close we think our result is to the real value If I wish to say *I have reason to believe with 90% confidence that μ = 2.45 ± 0.08 (n = 5)*, then the value 90% is referred to as the *confidence level* and 2.45 ± 0.08 is referred to as the *width of the confidence interval*. When computing confidence limits for the mean, we use the Student *t*-statistic:

$$\text{confidence limits for } \mu = \bar{x} \pm \frac{ts}{\sqrt{n}}$$

where *t* is the Student *t*-value, *s* the standard deviation and *n* the number of measurements. The value of *s* is found using the STDEV function. We determine *t* with the TINV function which has the syntax TINV(*probability, degrees of freedom*). The probability (α) equals (1 – the confidence level). For repeated measurements of the same object, the degrees of freedom (f) is given by $n - 1$.

We begin by finding the mean and confidence limits of a set of seven measurements. At the end of the exercise we will make the worksheet more flexible.

(a) On Sheet3 of Chap16.xlsx enter the text shown in columns A to D of Figure 16.5. Ignore columns F to I temporarily. Enter the values in column A.

(b) The formulas in column D are:

<table>
<tr><td>D2:</td><td>=AVERAGE(A3:A9)</td><td>Mean</td></tr>
<tr><td>D3:</td><td>=STDEV(A3:A9)</td><td>Standard deviation</td></tr>
<tr><td>D4:</td><td>=COUNT(A3:A9)</td><td>Number of measurements</td></tr>
<tr><td>D5:</td><td>=D3/SQRT(D4)</td><td>Standard error of the mean</td></tr>
<tr><td>D6:</td><td>=95%</td><td>Required level[1]</td></tr>
<tr><td>D7:</td><td>=TINV(1-D6, D4-1)</td><td>Student's t-statistic
(for $\alpha = 0.05, f = 6$)</td></tr>
<tr><td>D8:</td><td>=D7*D5</td><td>Confidence width</td></tr>
<tr><td>D11:</td><td>=D2</td><td>Mean, formatted to two places</td></tr>
<tr><td>D12:</td><td>=D8</td><td>Confidence limits (formatted[2])</td></tr>
</table>

[1] The value may be entered by typing 95%, or by typing 0.95 followed by formatting with the percentage tool.

[2] D12 is given a custom format of ±0.00. The symbol is produced with [Alt]+ 0177.

⊿	A	B	C	D	E	F	G	H	I
1	Mean, standard error and confidence limits								
2	data		mean	1.9543		data		mean	1.9760
3	1.93		stdev	0.0408		1.94		stdev	0.0369
4	1.92		n	7		1.99		n	10
5	2.02		stderror	0.0154		1.98		stderror	0.0117
6	1.97		conf level	95.0%		2.03		conf level	95.0%
7	1.98		t	2.4469		2.03		t	2.2622
8	1.96		conf width	0.0377		1.96		conf width	0.0264
9	1.90					1.95			
10			rounded values			1.96		rounded values	
11			mean	1.95		1.92		mean	1.98
12			conf width	± 0.04		2.00		conf limits	± 0.03

Figure 16.5

Some journals would require the data to be reported as having a mean of 1.95 and a standard error of 0.015 with $n = 7$. From this information the reader can compute the confidence limits for any required confidence level using the formula *confidence width = ±t × standard error*. Our worksheet allows the same. You may change the confidence level value in D6 to say 95.5, to find new confidence limits.

If we wish, we can simply compute the confidence limit with one formula:

=STDEV(A3:A9) * TINV(1 − D6, COUNT(A3:A9) − 1) / SQRT(COUNT(A3:A9)).

Complex formulas such as this, however, are error-prone.

Our worksheet would be useful for any experiment in which a measurement is repeated seven times or less. We can test this and at the same time double-check our worksheet.

(c) Use the Descriptive Statistics tool from the Data Analysis toolbox with the data in A3:A9. Do the values it reports for the mean, standard error, standard deviation and confidence limits agree with your worksheet? Erase the values in A3:A9 and enter three new values. Use the Descriptive Statistics tool again (you will recall that its values are static and you must re-run the tool after the data changes) and check for agreement.

Our worksheet will not give correct results with more than seven data items unless we make appropriate changes to all the formulas in column D that reference A3:A9. We can, however, make the worksheet flexible.

(d) Copy A2:D12 to F2. Modify the formulas in column I to read:
    I2:   =AVERAGE(F:F)
    I3:   =STDEV(F:F)
    I4:   =COUNT(F:F)
    Recall that the range references to F:F may be interpreted as F1:F1048576. This means the worksheet will give the correct result no matter how many values are entered. The empty cell in F1 and the text in F2 have no effect. Enter the values shown in F1:F12 and use Descriptive Statistics to validate your worksheet results.

(e) Save the workbook.

In Exercise 1 we saw that the CONFIDENCE function result does not agree with the results reported by the Descriptive Statistic tool. This tool always uses a t-value for an infinite value of f, the degrees of freedom; that is, it uses z-values. Its results may be acceptable when n is very large, or when it is known that the sample standard deviation (s) for the n measurements is always close to the population standard deviation (σ).

Exercise 4: The Experimental and Expected Mean

A series of measurements may be made on a specimen where there is an expected result. A chemist may analyze a chemical sample thought to be compound X and compare the results with the known composition of X to determine if the chemical sample is pure X. An engineer may measure the thickness of a metal plate and compare the results with the known thickness to test a new measuring device. Statisticians speak of hypothesis testing in these cases. For the chemist, we have the null hypothesis H_0: *The sample is compound X.* For the engineer, we have H_0: *This new instrument is suitable for the task.* Both may be stated as H_0: *This*

[3]Textbooks on statistics often use the terms *t(calculated)* and *t(table)*. We use the terms *t(experimental)* and *t(critical)* to avoid confusion since we 'calculate' both values.

The Excel Data Analysis Tools uses the terms *t(stat)* and *t(critical)*.

measured average ($\bar{x}$) *is the same as the expected average* (μ). There is the alternate hypothesis H_1: *This measured average* ($\bar{x}$) *differs from the expected average* (μ). Test the hypothesis on the mean by computing an experimental *t*-statistic and comparing it with the critical value[3] for a given confidence level. If the experimental *t* does not exceed the critical *t*, we dismiss the alternative hypothesis.

When testing hypotheses involving whether or not the mean value is a particular expected mean μ, the experimental *t* is computed using

$$t_{experimental} = \frac{|\bar{x} - \mu|}{s / \sqrt{n}}$$

where $\bar{x}$ is the mean and s is the standard deviation of the n measurements.

To calibrate a packing machine, an engineer has made a series of measurements of the raisin content in boxes of breakfast cereal. The required value is 33%. To test the results, we will construct a worksheet similar to that in Figure 16.6.

◢	A	B	C	D	E	F	G
1	Experimental and Expected Mean						
2							
3	data		Method 1			Method 2	
4	30.3		expected mean	33		expected mean	33
5	34.7		experiment mean	35.34		experiment mean	35.34
6	40.0		stdev	4.238		stdev	4.238
7	36.1		n	7		n	7
8	41.3		t expt	1.46		t expt	1.46
9	34.5		prob (required)	95%		alpha (required)	0.05
10	30.5		t(a, df)	2.45		prob (t, df, 2)	0.19
11			Fail to reject Null hypothesis			Fail to reject Null hypothesis	

Figure 16.6

(a) On Sheet4, enter the text shown in A1:C11 of Figure 16.6. For the time being, ignore the entries in columns F and G. Enter the experimental values in column A. Select A3:A13 and name A4:A13 as data. This will allow the worksheet to be used with up to 10 measurements, although we have only seven.

(b) The entries in column D are:
   D4:   33   Required mean
   D5:   =AVERAGE(data)   Calculated mean

D6: =STDEV(data)   Standard deviation
D7: =COUNT(data)   Number of measurements
D8: =ABS(D5–D4)/(D6/SQRT(D7))
                            Experimental *t*-value
D9: 95%   Required level of confidence
D10: =TINV(1 – D9, D7 – 1) Critical *t*-value
C11: =IF(D8>D10,"Reject Null hypothesis","Fail to reject Null hypothesis")

The entry in C11 is centered across C11 and D11 using the *Merge and Center* tool.

In this case the experimental *t*-value (1.46) does not exceed the critical value (2.45), so the null hypothesis is not rejected. With 95% certainty, we may say there is insufficient statistical evidence to believe the two values differ.

The alternative approach to this problem is to compute the probability that the mean value is statistically different from the expected value, that is, the *p*-value. We use TDIST to compute the *p*-value from the data and compare this to our required significance level (α), which is generally 0.05 or 5%. The syntax is TDIST(t, df, $tails$), where the first argument is our *t-expt*, df is the degrees of freedom, and *tails* has a value of 1 for a one-tail test or 2 for a two-tailed.

(c) Enter the text in F3:F10 (you could copy from column C and edit). Copy the formulas D4:D10 to G4 and make these changes:

G9: =1 – D9   The required α-value
G10: =TDIST(G8,G7 – 1,2)   The computed *p*-value
F11: =IF(G9<G10,"Fail to reject Null hypothesis","Reject Null hypothesis")

We fail to reject the null hypothesis, that the two means are shown to be statistically the same, since the calculated *p*-value (0.19) is greater than the stipulated α-value (0.05). We may interpret these results as saying that if the null hypothesis is true, there is a 19% probability that seven boxes taken at random will show a difference of 2.34 (=35.34 – 33.00) from the expected mean of 33. Note that we are saying that the difference between the found and expected means could occur by random errors. We are not necessarily saying this is an acceptable situation. The engineer may accept the accuracy of the machine but may decide to improve its precision in order to decrease the spread of the values.

In this problem we used a two-tailed t-value (the last argument in the formula) since we were concerned with both positive and negative differences from the expected mean. Consider another scenario: The packing machine fills, on average, 50 boxes a minute. After modification, 10 trials were made and the average filling rate was found to be 54.5 boxes/min with a standard deviation of 4.3. Has there been a statistically significant improvement in the machine? From these values t *expt* computes to 3.31 and a one-tailed p-value is 0.0035. So at the 5% level there has been a significant improvement, since the p-value is less than the α-value. In other words, we are not willing to accept the difference in the means as coming from random measurement errors.

Exercise 5: Pooled Standard Deviation

This Exercise is a prelude to the next one. The topic is repeated measurements on different specimens. We introduce the concept of the *pooled standard deviation* and the function DEVSQ. A biologist has measured the mercury content of seven fish taken from Lake Erie and obtained the results[4] shown in C4:H10 of Figure 16.7.

[4]This data was taken from D. A. Skoog and M. W. West, *Analytical Chemistry*, 2nd ed., p40, New York: Holt, Reinhart and Winston, 1974.

⊿	A	B	C	D	E	F	G	H	I	J
1	Pooled Standard Deviation									
2										
3	Sample	Replicates			Results				Mean	SSD
4	1	3	1.80	1.58	1.64				1.6733	0.02587
5	2	4	0.96	0.98	1.02	1.10			1.0150	0.01150
6	3	2	3.13	3.35					3.2400	0.02420
7	4	6	2.06	1.93	2.12	2.16	1.89	1.95	2.0183	0.06108
8	5	4	0.57	0.58	0.54	0.59			0.5700	0.00140
9	6	5	2.35	2.44	2.70	2.48	2.44		2.4820	0.06848
10	7	4	1.11	1.15	1.22	1.04			1.1300	0.01700
11	7	28							ss ->	0.20953
12										
13				mean of all measurements			1.67			
14				pooled standard deviation			0.10			

Figure 16.7

The pooled standard deviation is computed using the formula:

$$s_{pooled} = \sqrt{\frac{\sum SSD_i}{\sum (r_i - 1)}} \quad \text{or} \quad s_{pooled} = \sqrt{\frac{\sum SSD_i}{\sum r_i - n}}$$

where SSD_i is the sum of the squares of the deviations from the mean for the ith sample and r_i is the number of repeated measurements on the ith sample. The degrees of freedom are given in the divisors in these two equivalent formulas. Let us set up a worksheet to handle measurements of this type.

(a) On Sheet5 of Chap16.xlsx enter the text shown in the figure. Enter the values shown in A4:A10 and C4:H10.

(b) The number of samples n is found in A11 with the formula =COUNT(A4:A10). The number of repeated measurements for the first sample (r_1) is found in B4 with =COUNT(C4:H4) and this is copied down to B10. The total number of measurements $(\sum r_i)$ is found in B11 with =SUM(B4:B10).

(c) In I4 enter =AVERAGE(C4:H4) to find the mean of the first sample. The sum of the squares of the deviations from this mean (SSD_1) is found in J4 with =DEVSQ(C4:H4). These formulas are copied down to row 10. The sum of the SSD values is computed in J11 with =SUM(J4:J10).

(d) The mean for all measurements is given in G13 by =AVERAGE(C4:H10). The formula in G14 for pooled standard deviation is =SQRT(J11/(B11 − A11)).

(e) Save the workbook.

Exercise 6: Comparing Paired Arrays

In this Exercise we compare the mean from two sets of measurements made on a set of samples. Perhaps set A is the measurements using one technique, while set B was obtained from another. As in Exercise 4, we compute a standard deviation and use it to find a t-value. We compare the found t-value with the critical value computed for a specified α-value and the appropriate degrees of freedom. As before, we reject the null hypothesis for a two-tailed test if the experimental t-value is greater than the critical t-value.

For these circumstances (two measurements on several different samples) the t-value is computed using:

$$t_{\text{expt}} = \frac{\bar{d} - \mu_d}{s_d / \sqrt{n}}$$

where n is the number of paired measurements, $\bar{d}$ is the average of the differences between the pairs, μ_d is the expected average difference (usually 0), and

$$s_d = \sqrt{\frac{\sum (d_i - \bar{d})^2}{n-1}}$$

which is the standard deviation in the paired differences.

Our null hypothesis is, of course, that the mean of the differences between measurements A and B (the mean of column D) is statistically the same as the expected value of zero.

⊿	A	B	C	D	E	F	G	H	I	J	K
1	Paired sets										
2											
3		Data				Calculations			t-Test: Paired Two Sample for Means		
4	Specimen	A	B	diff		mean diff	0.16				
5	1	1.11	0.97	0.14		expected diff	0			A	B
6	2	3.77	4.33	-0.56		n	6		Mean	3.165	3.003
7	3	5.94	5.35	0.59		st dev	0.454		Variance	3.602	3.191
8	4	2.90	2.30	0.60		t expt	0.873		Observations	6	6
9	5	1.04	1.19	-0.15		alpha	0.05		Pearson Correlation	0.971	
10	6	4.23	3.88	0.35		t critical	2.571		Hypoth Mean Diff	0	
11									df	5	
12		Conclusion:		Fail to reject Null hypothesis					t Statistic	0.873	
13									P(T<=t) one-tail	0.211	
14						p approach			t Critical one-tail	2.015	
15						p from TTEST	0.423		P(T<=t) two-tail	0.423	
16		Conclusion:		Fail to reject Null hypothesis					t Critical two-tail	2.571	

Figure 16.8

In Exercise 4 we saw two methods to compare a measured mean with an expected value. We could call these the *t method* and the *p method*. We can also use a probability method for paired arrays of data. Without delving into the statistical theory, we will use the TTEST function which has the syntax: TTEST(*array1, array2, tails, type*) where *tails* has the same meaning as before and *type* is given a value of 1 for paired arrays.

(a) On Sheet6 of Chap16.xlsx, enter the text values shown in columns A:G of Figure 16.8. For now, ignore columns I:K. Enter the values shown in A5:C10.

(b) In D5 enter =B5–C5 and copy it down to row 10.

(c) The formulas in column G are:

    G4:   =AVERAGE(D5:D10)   Computes $\bar{d}$
    G5   0   The expected mean difference
    G6:   =COUNT(A5:A10)   Computes n
    G7:   =STDEV(D5:D10)   Computes s_d
    G8:   =G4*SQRT(G6)/G7   Compute t(*experimental*)
    G9:   0.05   The required α-value
    G10: =TINV(G9, G6-1)   Computes t(*critical*)
    G15: =TTEST(B5:B10,C5:C10,2,1) Computes the *p*-value
The formulas for the conclusions are:
    D12: =IF(G8<G10,"Fail to reject Null hypothesis",
                          "Reject Null hypothesis")
    D16: =IF(G15>G9,"Fail to reject Null hypothesis","Reject Null hypothesis")
These are each centered over four cells.

(d) Save the workbook.

We are led to the conclusion that the two methods give the same mean (with an α-value of 0.05) since (i) *t(experimental)* is less than *t(critical)* and (ii) the *p*-value computed by TTEST is greater than the α-value of 0.05.

To round off this Exercise, we use the *t-TEST: Paired Two Sample for Means* tool from the Data Analysis tool. The reader may wish to experiment or wait till the next exercise to see how to use this. Note that we should set the *Hypothesized mean difference* to 0 and the *alpha* value to 0.05 when completing the tool's dialog box. The results are shown in the figure. As expected, the results agree with our own calculations. The *t(experimental)* values in G8 and J12 are the same, as are the *p*-values in G15 and J15. These serve as useful checks but recall that the results from the tool are static whereas our calculations will be updated if new experimental array values are entered.

Exercise 7: Comparing Repeated Measurements

[5]When the two data sets are of equal size, this reduces to

$$s_p = \sqrt{(s_1^2 + s_2^2)/2} \ .$$

In the previous Exercise each specimen was measured once by each of two techniques. In this exercise the same specimen is measured repeatedly by two techniques. Our task is the same: To determine if the mean of the two sets of measurements is the same assuming equal variances. Once again, we have two statistical methods we could use: the *t* and the *p* methods. For the former we compute a pooled standard deviation using the formula[5]:

$$s_p = \sqrt{\frac{\sum_{set\,A}(x_i - \overline{x}_A)^2 + \sum_{set\,B}(x_j - \overline{x}_B)^2}{n_1 + n_2 - 2}} = \sqrt{\frac{s_1^2(n_1 - 1) + s_2^2(n_2 - 1)}{n_1 + n_2 - 2}}$$

from this we compute *t(experimental)* and compare it with *t(critical)*. The experimental *t*-value is found using:

$$t_{experimental} = \frac{\overline{x}_1 - \overline{x}_2}{s_p}\sqrt{\frac{n_1 n_2}{n_1 + n_2}} = \frac{\overline{x}_1 - \overline{x}_2}{s_p\sqrt{(1/n_1 + 1/n_2)}}$$

For the *p* method we will again use the Microsoft Excel functions TDIST or TTEST to find a probability value, which we will compare to the required α-value. We will also use the Data Analysis tool *t-Test: Two-Sample Assuming Equal Variance* to check our results.

(a)   On Sheet7 of Chap16.xlsx enter the text shown in A1:D19 of Figure 16.9. Enter the experimental values in columns A and B. Name A5:A19 as **A** and B5:B19 as **B**. This will allow the worksheet to be used with up to 15 data points when it is used with other data.

◢	A	B	C	D	E	F	G	H	I	J
1	Comparing repeated measurements									
2										
3		Data			Calculations			t-Test: Two-Sample Assuming Equal Variances		
4	A	B			A	B				
5	179.738	179.864		mean	179.7285	179.6645			A	B
6	179.707	179.611		st dev	0.016	0.144		Mean	179.7285	179.6645
7	179.731	179.537		ssd	1.852E-03	1.452E-01		Variance	0.0002646	0.0207469
8	179.722	179.903		n	8	8		Observations	8	8
9	179.745	179.543		s pooled	0.1025			Pooled Variance	0.0105057	
10	179.731	179.661		t expt	1.249			Hypothesized Mean Di	0	
11	179.749	179.544		prob	95%			df	14	
12	179.705	179.653		df	14			t Stat	1.2488123	
13				t theory	2.145			P(T<=t) one-tail	0.1161064	
14				outcome	Fail to reject Null Hypothesis			t Critical one-tail	1.7613101	
15								P(T<=t) two-tail	0.2322128	
16					p method			t Critical two-tail	2.1447867	
17				p expt	0.232					
18				t-test	0.232					
19				alpha	0.05					
20				outcome	Fail to reject Null Hypothesis					

Figure 16.9

The entries in E9:14 and E17:E20 are each merged and centered over the next column. Cells E10 and E14 have *Wrap Text* turned on (in the *Alignment* group on the *Home* tab), and these rows have increased heights.

(b) The formulas in columns E and F are:

E5: =AVERAGE(A)
E6: =STDEV(A)
E7: =DEVSQ(A)
E8: =COUNT(A)
F5: =AVERAGE(B)
F6: =STDEV(B)
F7: =DEVSQ(B)
F8: =COUNT(B)
E9: =SQRT((E7+F7)/(E8+F8-2))
E10: =(ABS(E5 - F5)/E9) * SQRT((E8*F8)/(E8+F8))
E11: 95%   The required confidence level
E12: =E8 + F8 - 2   The degrees of freedom
E13: =TINV(1-E11,E12)
E14: =IF(E10<E13,"Fail to reject Null Hypothesis","Reject Null Hypothesis")

Comparing the t(experimental) value of 1.249 in E10 with the t(critical) value of 2.145 in E13, we fail to reject the null hypothesis that the two means are statistically the same.

(c) For the p method, the formula in E17 is =TDIST(E10,E12,2) and in E18 it is =TTEST(A, B, 2, 2) for a two-tailed test with sets having equal population variances. In E19 we use =1-E11 to compute the required alpha. It is left to the reader to compose the formula in E20.

The results here lead to the same conclusion: that the null hypothesis cannot be dismissed.

You may wonder why we used two formulas for the *p* method. The simple answer is that TTEST is only of use when the two arrays are of equal size. The longer method, which involves computing a *t*-value from which to compute the *p*-value, is applicable when the sets are of unequal size.

(d) Use *Data / Data Analysis* and select the tool *t-Test: Two-Sample Assuming Equal Variance.* Complete the dialog as shown in Figure 16.10. The two *t*-statistics from the tool agree with our calculations, and so do the *p*-values.

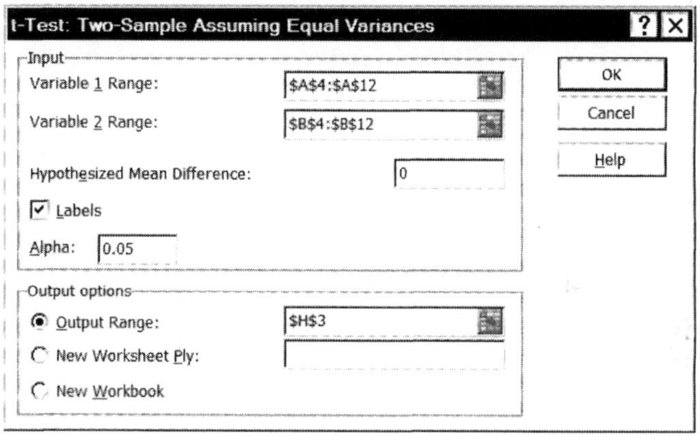

Figure 16.10

Unlike TTEST, this tool may be used with arrays of unequal size. We can also test if the means differ by a specified nonzero amount by entering a value in the hypothetical mean difference box. If we wish to do a similar test with formulas, the *t*(experimental) value must be computed using:

$$t_{experimental} = \frac{(\bar{x}_1 - \bar{x}_2) - (\mu_1 - \mu_2)}{s_p} \sqrt{\frac{n_1 n_2}{n_1 + n_2}}$$

where $(\mu_1 - \mu_2)$ represents the hypothesized difference in the population means.

Exercise 8: The Calibration Curve Revisited

In Chapter 8 we saw how to chart a calibration curve and add a trendline. We also used the functions SLOPE, INTERCEPT and LINEST to find the slope and intercept of the line of best fit. This line, of course, has uncertainties associated with it. The LINEST function not only gives us the values for the slope and intercept, it also gives the errors associated with them. Let s_b be the standard error (uncertainty) for the intercept b, s_m the standard error for the slope m, and s_y the standard error for the estimate of y. If $y*$ is the measured signal for an unknown, then the value of the unknown is computed using

$$x* = \frac{y*(\pm s_y) - b(\pm s_b)}{m(\pm s_m)}$$

⊿	A	B	C	D	E	F
1	Uncertainty in a Calibration Curve					
2						
3	x	y			Linest	
4	1	2.86	slope	2.288	0.568	intercept
5	2	5.20	error in slope	0.030022	0.099572	error in intercept
6	3	7.40	R²	0.999484	0.094939	error in estimate of y
7	4	9.60	F statistic	5808	3	degrees of freedom
8	5	12.10	regression ss	52.34944	0.02704	residual sum of squares
9						
10						
11		Index with Linest			value	err	%err
12	m	2.288	y*	6.55		
13	b	0.568	numerator	5.982	0.1376	2.30%
14	sm	0.0300	denominator	2.288	0.0300	1.31%
15	sb	0.0996	x*	2.615	0.0692	2.65%
16	sy	0.0949				

Figure 16.11

In this Exercise we make a calibration curve and determine $x*$ for a measured $y*$ using the equation above. The function LINEST is used to find the required parameters. We will see how a combination of INDEX and LINEST allows us to generate only those parameters that are necessary for the task. We shall need to recall that errors are combined using $e_3 = \sqrt{e_1^2 + e_2^2}$ and that for multiplication and division, we must work with percentage errors.

(a) On Sheet8 of Chap16.xlsx enter the text shown in Figure 16.11.

(b) Enter the calibration data in A4:B8. Name the columns as x and y, respectively.

(c) Select D4:E8, enter the formula =LINEST(y, x, TRUE, TRUE), and press ⌃Ctrl+⇧Shift+Enter↵ to complete the array formula. The entry will appear in the formula bar surrounded by braces {} because it is an array formula.

(d) To see how we may obtain certain parameters from the LINEST function, enter the formulas shown here. These are not array formulas, so complete them normally.

B12: =INDEX(LINEST(y, x, TRUE, TRUE), 1,1)
B13: =INDEX(LINEST(y, x, TRUE, TRUE), 1,2)
B14: =INDEX(LINEST(y, x, TRUE, TRUE), 2,1)
B15: =INDEX(LINEST(y, x, TRUE, TRUE), 2,2)
B16: =INDEX(LINEST(y, x, TRUE, TRUE), 3,2)

The first formula returns the LINEST value that would normally be in the first row and first column, that is, the slope of the line of best fit. Likewise, the second gives us the intercept, which is in row 1, column 2, of the LINEST array.

(e) Name the cells in B12:B16 with the text to their left. This will make it easier to understand the formulas that follow.

(f) For the purpose of the Exercise, assume our measured signal had a value of 6.55. Enter this value in D12. Enter the following formulas:

D13: =D12 - b   The numerator $(y^* - b)$
E13: =SQRT(sy^2+sb^2) The error in the nominator
F13: =E13/D13   The percentage error in the nominator
D14: =m   The denominator m
E14: =sm   The error in the denominator
F14: =E14/D14   The percentage error in the denominator
D15: =D13/D14   The value $x^* = (y^* - b)/m$
E15: =D15*F15   The error in x^*. This will mean nothing until F15 is computed
F15: =SQRT(F13^2 + F14^2)
                       The percentage error in x^*

When using a spreadsheet (or a calculator) to do such computations, we let it use its full precision. We may wish to format the cell to show a limited number of digits if the spreadsheet is to be displayed to others. We must round off the values when reporting the results. We would report x^* as $2.59_3 \pm 0.07_0$ or $2.59_3 \pm 2._7\%$.

Exercise 9: More on the Calibration Curve

[6]See, for example, P. C. Meier and R. E. Zünd, *Statistical Methods in Analytical Chemistry*, Wiley, New York, 1993.

[7]AMC Technical Brief No. 22, ed. M. Thompson, March 2006, Royal Society of Chemistry, London (http://www.rsc.org/images/Brief22_tcm18-51117.pdf). Reproduced by permission of the Royal Society of Chemistry.

The statistical analysis in the previous Exercise ignores the fact that the estimations of the slope and intercept are interdependent. A full treatment of the alternative approach is beyond the scope of this book. The interested reader may wish to consult an advanced statistics book.[6] We will take a more pragmatic approach and give the *how* without the *why*. The author is indebted to the Royal Society of Chemistry for permission to quote from one of its technical briefs.[7] Be aware that this brief uses $y = a + bx$ as the equation of a straight line; we shall stay with $y = mx + c$.

The regression line has an associated confidence interval. Figure 16.12 show the 95% confidence limits for some data with an R^2 value of only 0.58. Poor data was used to enable us to clearly see the three lines. The expression for the confidence interval for the computed Y- values is:

$$CI(Y) = \pm t(\alpha, df) \cdot S_{yx} \cdot \sqrt{\frac{1}{n} + \frac{(x_i - \bar{x})^2}{S_{xx}}}$$

This is for a set of n data pairs having an average x-value of $\bar{x}$. We will see later how S_{yx} and S_{xx} may be computed in Excel.

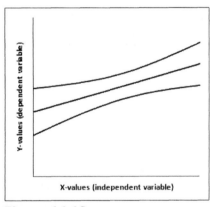

Figure 16.12

In general, regression tools and charts are used to predict a value of a dependent variable (y) from a measured value of an independent variable (x).

If x^* is the new x-value, then the confidence interval for the predicted y^*-value is found using:

$$CI(y^*) = \pm t(\alpha, df) \cdot S_{yx} \cdot \sqrt{\frac{1}{n} + \frac{(x^* - \bar{x})^2}{S_{xx}}}$$

When we use a calibration curve we reverse matters. We take a new y^* value and estimate its x^* value. This type of analysis is sometimes called inverse regression. We may use either of these equations to compute the confidence limits for reverse regression:

$$CI(x^*) = \pm t(\alpha, df) \cdot \frac{S_{yx}}{|m|} \cdot \sqrt{\frac{1}{n} + \frac{1}{k} + \frac{(y^* - \bar{y})^2}{m^2 \cdot S_{xx}}}$$

$$CI(x^*) = \pm t(\alpha, df) \cdot \frac{S_{yx}}{|m|} \cdot \sqrt{\frac{1}{n} + \frac{1}{k} + \frac{(x^* - \bar{x})^2}{S_{xx}}}$$

In these equations: m is the slope of the fit and n the number of points in the regression data, while k is the number of duplicates in the y^* measurements. The quantity S_{yx} (the standard error of the estimate) may be found with the Excel STEYX function while S_{xx} (sum of the squares of x deviations) is computed with the DEVSQ function. The approximations inherent in these equations are valid only when:

$$\frac{t(\alpha, df) \cdot S_{yx}^2}{m^2 \cdot S_{xx}} \leq 0.05$$

For our example we have calibrations consisting of five data pairs (A5:B9 in Figure 16.13). In F4:F13 we find the slope and intercept of the line of best fit together with various quantities needed to compute the confidence levels. In B11:C11 we test to see if our data meets the criterion (equation above) to allow us to use the CL formulas. Then in F15:F20 we do the actual CL calculation; note how the two formulas give the same results. The reader may decide not to enter the documentation in column G and rows 22:23. Because we have somewhat involved formulas, this is a most appropriate time for using named cells. As we have planned the work in advance we will actually name the cells before they contain any data or formulas.

(a) Open Chap16.xlsx and on Sheet 9 enter the text shown for A1:F20 in Figure 16.13.

(b) Name A5:A9 as X and B5:B9 as Y.

(c) Select E4:F17 and use the naming tool to name the cells F4:F17.

	A	B	C	D	E	F	G
1	Calibration Curve Uncertainty (Confidence Limits)						
2							
3	Calibration data				Parameters		Formulas
4	x	y			m	2.353	=SLOPE(y,x)
5	1.12	2.86			c	0.289	=INTERCEPT(y,x)
6	2.05	5.20			n	5.000	=COUNT(x)
7	2.99	7.40			df	3	=n-2
8	4.03	9.60			Syx	0.131	=STEYX(y,x)
9	4.99	12.10			SSx	9.452	=DEVSQ(x)
10					avgx	3.036	=AVERAGE(x)
11	Test	0.0033	TRUE		avgy	7.432	=AVERAGE(y)
12					p	0.950	0.95
13					t	3.182	=TINV(1-p,df)
14	Measured y values for unknown						
15	YY				k	5	=COUNT(YY)
16	6.55				avgYY	6.550	=AVERAGE(YY)
17	6.47				compXX	2.661	=(avgYY-c_)/m
18	6.56				CL x*	0.114	see below
19	6.57				CL x*	0.114	see below
20	6.60				XX*	2.66± 0.11	=ROUND(compXX,2)&"± "&ROUND(F18,2)
21							
22		CL x*	=t*(Syx/ABS(m))*SQRT(1/n+1/k+(avgYY-avgy)^2/(m^2*SSx))				
23		CL x*	=t*(Syx/ABS(m))*SQRT(1/n+1/k+(compXX-avgx)^2/SSx)				

Sheet6 (3) / Sheet7 / Sheet7 (2) / Sheet8 / Sheet9(2)

Figure 16.13

(d) Using the information shown in column G in the figure, enter formulas in F4:F13. If you use the pointing method, Excel will use the cell names in the formulas. This will make it much easier to check that your formulas are correct.

(e) Now we will see if we satisfy the criterion that allows the CL values to be computed with the approximate equations. In B11 enter =t^2*Syx^2/(m^2*SSx) and in C11 enter =B11<0.05. The result of TRUE tells us we may proceed.

We have a calibration. Now we can enter the data for the "unknown." For this demonstration we assume the experimenter has repeated his measurements five times on the same sample.

(f) Enter the five readings in A16:A20 and give the range the name YY.

(g) Enter the formulas in F15:F20. Save the workbook.

The formula in F15 gives us the *x**-value. In F16 and F17 we compute CL using each of the equations shown at the start of the exercise. Finally in F20 we use a formula to neatly display the *x**-value with its confidence limits.

The companion website contains an Excel file called *CalibrationCurve.xlsx,* which has worksheets showing the use of Excel with the data from the Royal Society of Chemistry technical brief and a demonstration of the use of a spinner coupled to a calibration curve.

Problems

[8]W. J. Youden, *Experimentation and Measurement*, U.S. Department of Commerce, 1984.

1. The thickness of two paper samples was measured four times for each sample.[8] The results for Sample A were 772, 759, 795, and 790 (the units are inches × 10^{-4}). For Sample B they were 765, 750, 724, 753. Does this data suggest the samples have the same or different thickness?

2. *Is there a statistical difference in data sets A and B?

A	B
2.31017	2.30143
2.30986	2.29890
2.31010	2.29816
2.31001	2.30182
2.31024	2.29869
2.31010	2.29940
2.31028	2.29849
	2.29889

[9]W. Mendelhall et al., *Statistics for Management and Economics*, Duxbury, Belmont, CA, 1993.

3. *To test for any difference in wear, 10 tires (5 of type X and 5 of type Y) were randomly placed on the front rims of five cars and wear measurements taken after the cars had been driven a set number of miles; see the following table.[9] Use a Data Analysis tool to statistically decide if the two means of the two tire wear values differ.

	Tire Type	
Automobile	X	Y
1	10.6	10.2
2	9.8	9.4
3	12.3	11.8
4	9.7	9.1
5	8.8	8.3

4. Grissom and Sara have analyzed samples from the same crime scene. Each has reported the mean values of their tests. Also, in accord with CIS policy, they have reported how many tests were made and the SSD (sum of squares of deviations from the mean) values. From the data in the following table can you state there is a statistical difference in their results?

	Grissom	Sara
Mean	59.15	59.62
Number of analyses	6	4
SSD	0.0214	0.0295

5. *A widget manufacturer measured the lifetime of 5000 of his product. The data is approximately normally distributed with an average of 585 hours with a standard deviation of 89 hours. If he plans to sell two million and promises to give buyers a free widget if theirs lasts less than 750 hours, how many should be make? How many are expected to last between 600 and 700 hours?

[10] G. Keller et al. *Statistics for Management and Economics,* Duxbury, Belmont, CA, 1984.

6. The lifetimes[10] for Acme Premium tires are approximately normally distributed with a mean of 45,000 miles and a standard deviation of 2500 miles. The tires have a 40,000 mile warranty. (i) What percentage of the tires will fail before the warranty expires? (ii) What percentage of the tires will fail within 1000 miles of the warranty expiration?

7. Make a duplicate of Sheet2 by dragging its tab to the right giving a worksheet called Sheet2(2). Answer Yes to all questions about keeping cell names. Change the formula in E5 to =NORMDIST(C5,mean,stdev,FALSE). Copy this down to E17. The chart is disappointing—we have not finished. We need to "normalize" the histogram data. Select D2:E17 and change the array formula to read =FREQUENCY(data,C5:C17)*($E$18/COUNT(data). Note that the SUM values in row 18 are now the same. The result is similar to that in Exercise 2, but the peak of the bell curve is now displaced half an increment to the left. Change E5 to =NORMDIST(C5-0.125,mean,stdev,FALSE) and copy down the column.

8. Apply the method of Exercise 9 to the data in Exercise 8 to see the difference in the computed confidence intervals.

[11]F. D Snell et al., *Colorimetric Methods of Analysis*, Van Nostrand, New York, 1951.

9. The accompanying table[11] shows the calibration data for the chemical analysis of silica; X is the known amount of silica and Y is the absorbance of the solution. An unknown sample had an absorbance value of 0.242. Find the amount of silica in the unknown and give the 95% confidence limits.

X	0.000	0.020	0.040	0.060	0.080	0.100	0.120
Y	0.032	0.135	0.185	0.268	0.359	0.435	0.511

10. We have not had time to consider the F-distribution. Using information from a textbook and/or the Internet, look again at the data in Problem 4 and determine if there is any statistical reason to think there is a difference in the precision of the two agents.

11. Write a UDF to compare the mean of two ranges at the 0.05 significance level and return either *Fail to reject Null hypothesis* or *Reject Null hypothesis*, depending on the relative values of *t-expt* and *t-critical*. The UDF should be called with something like =TwoMeans(A1:A10, B1:B15).

12. Write a UDF to compare the mean of a range with an expected value and return either *Fail to reject Null hypothesis* or *Reject Null hypothesis* depending on how the computed *p*-value compares to 0.05. The UDF should be called with something like =ExpectedMean(A2:A11, B2, 0.05). Test your function by comparing this data with an expected value of 59.3 at a significance of 0.05.

59.09	59.17	59.27	59.13	59.10	59.14

Report Writing

Documentation

If your worksheet will be used over a long period of time by you or others, then documentation is essential. In the first part of this chapter, we look at a variety of tools that can be used for this purpose. For this discussion we will consider a worksheet similar to that depicted in Figure 17.1.

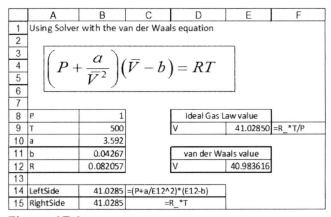

Figure 17.1

The topics we will cover are:
- How to go from worksheet to picture.
- How to show the formulas in F9 and C14.
- How to construct the equation.
- Using a screen capture application.
- Using the Excel add-in called MathLook.

You may wish to create the worksheet above (without the equation right now) in a workbook to be saved as Chap17.xlsx so you can experiment.

Picture of Worksheet

These are the steps for making Figure 17.1.
(i) In *Page Layout | Sheet Options* check the box *Headings | Print*. This will give us the row and column headings seen in the figure.
(ii) Select the range to be captured. Use the command *Home | Paste | As Picture*. This is somewhat confusing since we are copying, not pasting! In the next menu select *Copy as Picture*.

(iii) In the next dialog select *As Printed* to ensure we get the heading.

(iv) Move to where the picture is required (this could be on the same worksheet) and use the normal paste command.

Display Formulas

As mentioned before, the short cut [Ctrl]+` (where ` is the key to the left of 1 on the top row of the keyboard) causes the worksheet to display formulas not values. The shortcut is a toggle—use it the second time to return to values.

To "capture" a cell's formula for display in another cell or in another application such as Word:

(i) With the source cell selected, use the mouse to highlight the entire formula in the formula bar. Use the [Ctrl]+C shortcut to copy it to the Clipboard.

(ii) Press [Esc] key. This is most important; otherwise the cell formula will be compromised.

(iii) Move to where the formula is required and use the normal paste command. If the target is an Excel cell, you must precede it with an apostrophe to have it display as text. Otherwise you will have just copied the formula but with no reference adjustment, which can sometimes be useful.

Note that this is a static procedure, meaning that a change in the source cell will not result in a change in the displayed version. For a quicker and dynamic solution, one can use the following UDF.

```
Function showform(mycell)
      showform = mycell.Formula
End Function
```

If this is placed in the *Personal.xlsb* file, it will be available in all other workbooks. It would be found in the *Insert Function* dialog in the User Defined category as Personal.xlsb!Showform.

Exercise 1: Creating an Equation

$$\int_0^\pi \frac{1}{x^2}dx$$

Microsoft provides an applet (a small application that is run from within another application) called Equation Editor, which may be used in programs such as Word or Excel to create an equation. With a little practice and experimentation you will be able to create complex equations. In this Exercise we create the expression at the left to get you started.

(a) On Sheet2 of Chap17.xlsx, use the command *Insert | Text | Object* and select *Microsoft Equation 3.0*. Figure 17.2 shows how the worksheet appears. Note that the row of commands

(File, Edit, View...) relates to the Equation applet not to Excel. You may need to use the View command to have the floating toolbar visible.

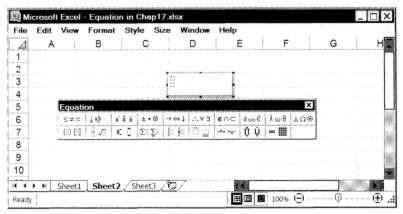

Figure 17.2

(b) To draw the integral sign, click the mouse pointer over the fifth item on the bottom row of the Equation Editor toolbar. Move the pointer to the second item on the top row of the drop-down menu, since we need an integral sign with two limits.

(c) Experiment by tapping the [Tab↹] key and with holding down the [⇧ Shift] key while tapping [Tab↹]. The L shape that moves around is the *insertion point*. When a box has something typed in it, the L is reversed. Now use the mouse to move the insertion point to the box which will hold the lower limit. In this box type 0.

(d) Using either the mouse or [Tab↹], move to the box where the upper limit will go. Open the ninth item on the top row of the toolbar and click on the π symbol.

(e) Use [Tab↹] to move the insertion point into the box at the right of the integral sign. Move the pointer to the second item on the bottom row of the toolbar and select the first item on the top row of the drop-down menu—two open boxes stacked vertically with a bar between them. We need this template for the $1/x^2$ part of the expression.

(f) Move the insertion point to the top box and type 1. Move the insertion point to the bottom box and type x. Experiment using the mouse to relocate the insertion point.

(g) We need an object to hold the superscript. Move to the third item on the bottom row of the toolbar and select the first item from the menu. You should now have a superscript box in which to type the 2. Alternatively, type the 2, select it, and now open the template for a superscript.

(h) Press `Tab⇄` to move the insertion point to the far right of the equation. Type dx.

(i) Click the mouse anywhere outside the equation box to close the Equation Editor applet.

(j) If you right click an Equation object and select *Equation Object | Open* you can work in a new window with more space (Figure 17.3). You may need to use *View | Toolbar*. Sometimes the author begins an equation, closes, and uses this technique to complete the task.

Figure 17.3

As you may have discovered, you cannot use `Spacebar` when forming an equation—the applet looks after the spacing of items. You can however, use `Ctrl`+`⇧ Shift` to add more spacing.

By default the Equation Editor uses italics for variables such as x and regular font for digits and anything it thinks is a function such as Exp or Ln. You can enter normal text, including spaces, in an equation box by using the Text item in the Style menu.

Some users have reported having better success if they compose the equation in Word and copy the object to the Excel worksheet. In Word you use the same command *Insert | Text | Object.* and select *Microsoft Equation 3.0.* The simple *Insert | Symbol | Equation* generates an in-line equation. It also has some 'canned' equations. Equations made this way may also be pasted into Excel. Be careful when resizing an equation object to pull a corner handle, not a side handle; otherwise you will distort the equation.

Now experiment to make the van der Waals equation as in Figure 17.1.

Screen Captures

Earlier we saw how to use *Copy As Picture* to get a snapshot of a worksheet range. A screen capture can be more useful. Try this with Sheet1 as the active worksheet: Press [PrtScr], move to a new Word document, and press [Ctrl]+V. You have a picture of the entire screen. A graphics program would be needed to crop the picture to select the important area. There are many screen capture programs that will more readily produce better quality captures with much more flexibility. One of the better known free applications is IrfanView from a website of the same name. The author's favorite is SnagIt from www.TechSmith.com. Many of the figures in this book were made using it including Figure 17.4, which shows the worksheet of Figure 17.1.

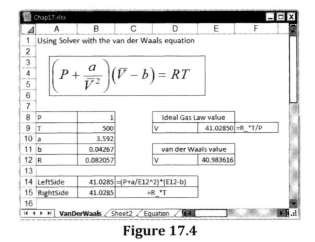

Figure 17.4

MathLook™

MathLook from www.uts.com is an Excel add-in that generates graphics from the formulas in cells. While they have the drawback of being static, they are marvelous for checking your worksheet against the mathematical model being used. The formulas are not just pictures of Excel formulas but, where appropriate, they more closely resemble mathematical formulas. Figure 17.5 was generated from the worksheet shown in Figure 16.13.

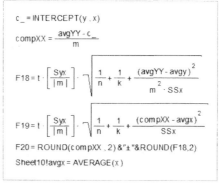

Figure 17.5

The first formula is plain Excel, but the formula in F18 which was =t*(Syx/ABS(m))*SQRT(1/n+1/k+(compXX-avgx)^2/SSx) is rendered in a much more meaningful way. Note the last formula; MathLook used *Sheet10!avgx* because there are other cells with the name *avgx* on other sheets. Figure 17.6 was generated from the worksheet in Figure 17.4. MathLook is aware that B14 and B15 were used in a Solver constraint!

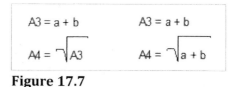

Figure 17.6

Let a worksheet have numbers in A1 and A2, named *a* and *b*, respectively. Let A3 have the formula = a + b and A4 have =SQRT(A3). By changing a setting called "current depth," I can have these rendered in either of the two ways shown in Figure 17.7.

A3 = a + b   A3 = a + b

$$A4 = \sqrt{A3} \qquad A4 = \sqrt{a+b}$$

Figure 17.7

In addition, MathLook has a tab for managing range names. At its simplest it gives a neat overview of the names on the current worksheet. But it can also be used to give cells names. If I used

MathLook to name A3 as *plus* in the last example, then all cells would become aware of this and change any reference to *A3* to *plus*. In this way the cell A4 would get the formula =SQRT(plus). Excel requires an extra step for the cells to inherit new names.

Copy and Paste or OLE?

In the remainder of this chapter, we learn how to place Microsoft Excel workbook data and charts into a word processor document. There are two very different ways to do this: (i) Using copy and paste, or (ii) with Object Linking and Embedding (OLE). We will examine these methods in detail in the exercises. The algorithm below will help you choose the appropriate method.

Are you sure that the workbook is complete and the report will never need updating?
  Yes: Use copy and paste.
  No: Will you always have access to the workbook?
            Yes: Use linking.
            No: Use embedding.

Exercise 2: Copy and Paste

The method described in the Exercise will work with all Microsoft Office products, such as Word and PowerPoint. They also work with many other applications. If a simple Paste does not give the required result, look in the Paste Special options.

In this Exercise, we will copy data and a chart from an Excel workbook to a document you are writing with a word processor application such as Microsoft® Word or Corel® WordPerfect. We need a simple workbook with data and a chart. Let us assume we have run an experiment to find the value of a resistor by measuring the currents passing through the resistor when various voltages are applied. Since $I = V/R$, the slope of a plot of I vs V will be $1/R$.

(a) Open Chap17.xlsx and on Sheet 2 enter the data shown in A1:B11 of Figure 17.1. The cell B11 contains the formula =1000/SLOPE(B4:B9,A4:A9) where the factor of 1000 accounts for the fact that the current was measured in milliamps.

(b) Construct a chart similar to that in Figure 17.8. Insert a trendline without the formula being displayed.

(c) Save the workbook as Chap17.xlsx.

(d) Without closing Excel, open your word processor. Put some text into a new document as in Figure 17.2 but without the table or chart.

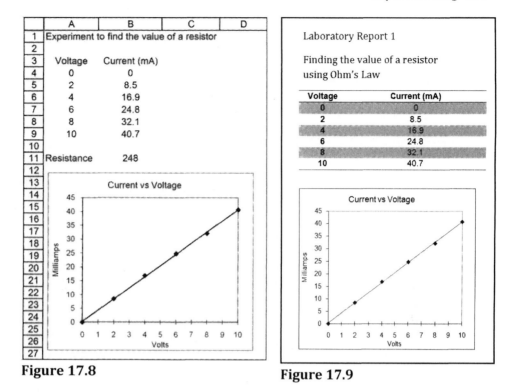

	A	B	C	D
1	Experiment to find the value of a resistor			
2				
3	Voltage	Current (mA)		
4	0	0		
5	2	8.5		
6	4	16.9		
7	6	24.8		
8	8	32.1		
9	10	40.7		
10				
11	Resistance	248		

Laboratory Report 1

Finding the value of a resistor
using Ohm's Law

Voltage	Current (mA)
0	0
2	8.5
4	16.9
6	24.8
8	32.1
10	40.7

Figure 17.8

Figure 17.9

(e) Make Excel the active application. Select the range A3:B9 in Chap17.xlsx and use the shortcut [Ctrl]+ C to copy it.

(f) Make Word 2007 active. Use the shortcut [Ctrl]+ V. The data from the Excel worksheet is inserted into the document as a table without borders (Figure 17.9). Note that the table uses the same font as the worksheet but Word displays a Paste Options smart tag (Figure 17.10) that gives a quick way to change that. The same tag has an option to paste as just text without the table structure.

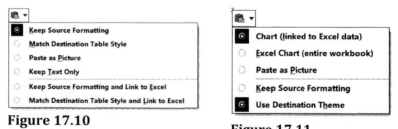

Figure 17.10

Figure 17.11

(g) Word's *Table Tools | Design | Table Style Options* gives a quick way to add border and/or shading to the table.

(h) Return to the workbook. Right click the Chart Area, and use the Copy item in shortcut menu. In Word, right click where the chart is to go and use Paste. The chart may not be in the right place but proceed to the next step before dragging it.

(i) By default, the pasted chart will be linked to the original Excel worksheet. If you open the Paste Option smart tag (it will disappear if you do anything else after pasting), you will see various options (Figure 17.11). We will leave it at the default. We say the chart is *linked* to the Excel file.

(j) Drag the chart to the correct position. Be careful to drag a chart area, not a chart element such as the title.

(k) In Microsoft Office 2007, the chart engine is common to all applications. This means you can modify the chart in Word in the same way you do in Excel. Formatting or changing the chart type does not affect the original chart. However, if the data is changed in the original Excel chart, the Word chart also changes. If in the Word chart you right click and select *Edit Data,* then the Excel file is opened for editing.

(l) If you want the chart to be totally independent of the Excel file, you can either use the Picture option on the smart tag or use *Paste Special* rather than *Paste* and select one of the picture options.

Exercise 3: Object Embedding

To get the feel for embedding, we shall embed (rather than link) the same chart as in Exercise 1 in a new Word document.

(a) In the Chap17.xlsx workbook click once on the chart. Click the Copy button.

(b) Move to your word processor and start a new document. Right click anywhere and select Paste.

(c) Open the Paste smart tag (see Figure 17.11) and select *Excel Chart (entire workbook)*. We have embedded an Excel document within a Word Document.

Because all Microsoft Office applications share the same chart engine, you can edit the chart directly in Word. If you right click the chart and select Edit Data, then an Excel window opens and the tile bar reads *Chart in Microsoft Office Word - Microsoft Excel.* This file is entirely unconnected with your original Chap17.xlsx file. If you save the word processing document, a copy of the workbook is saved with it, not as a separate file but as part of the Word document file. You could give a copy of the Word file to a colleague who could then modify both the Word document and the Excel workbook, provided their computer had Office installed.

Microsoft Visio®

The Microsoft Visio application is a useful tool for making simple drawings. It is not a CAD application but has some very powerful features. It would not be appropriate to try to give any instructions here, but a screen shot (Figure 17.12) may whet your appetite. Visio was used for the flow chart in Figure 10.8.

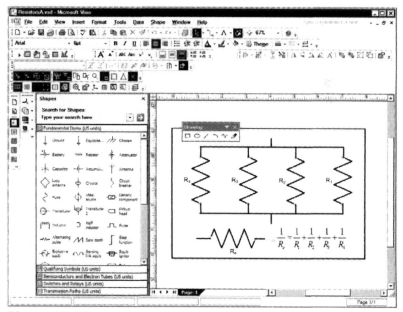

Figure 17.12

Answers

Chapter 2

4. Some possible improvements include:
   - Unlocking the input cells, then protecting the worksheet so that formulas are not accidentally changed or deleted. See Exercise 5 in Chapter 5.
   - Having output cells not display zeros. This can be done with custom formatting. Alternatively, use *Office* to open the *Excel Options* dialog, locate the *Advanced* tab and in *Display options for this worksheet*, uncheck the box *Show zeros in cells that have zero values.*
   - Using the ROUND function to have integer values displayed for *Packages Needed.*
   - Write VBA code (Chapter 10) that would clear all the input cells when a control on the worksheet was clicked.

5. The formulas are shown in the following table. The factor 24 converts time in days to time in hours.

	D	E	G	H
5	=A5-A4	=B5-B4	=E5/(D5*24)	=(B5-B4)/((A5-A4)*24)

Chapter 4

1. The required formulas are as follows.

B7	=B3*(C6-B6)/(C5-B5)
C7	=$B$3*(D6-B6)/(D5-B5)
D7	=$B$3*(E6-C6)/(E5-C5)
E7	=$B$3*(F6-D6)/(F5-D5)
F7	=$B$3*(G6-E6)/(G5-E5)
G7	=B3*(G6-F6)(G5-F5)

4. The formulas are:

n	=COUNT(B3:G3)
Σ(xy)	=SUMPRODUCT(B3:G3,B4:G4)
Σ(x)	=SUM(B3:G3)
Σ(y)	=SUM(B4:G4)

$\Sigma(x^2)$	=SUMSQ(B3:G3)
slope	=(B6*B7-B8*B9)/(B6*B10-B8^2)
intercept	=(B9-E7*B8)/B6

5. The formulas are as follows.

Term 1	=0.5*PI()*B4^2	
Term 2	=B4^2*ASIN(B5/B4)	
Term 3	=B5*SQRT(B4^2-B5^2)	
Volume	=B3*(E3-E4-E5)	ft^3
Volume	=E6*7.48051945	gallons US

Chapter 5

2. These give the required result:

4	=MATCH(B4,frame,0)+1
7	=MATCH(B3*12,height,0)
161	=INDEX(F2:H16,MATCH(B3*12,height,1), MATCH(B4,frame,0))

3 through 5. Sample data and formulas are as follows.

	A	B	C	D
4	-6	23	Sum those > 0	=SUMIF(A1:A10,">0")
5	7	113	Sum^2 those > 0	=SUMPRODUCT(--(A1:A10>0),(A1:A10)^2)
6	-10	113	Sum^2 those > 0	=SUMSQ(IF(A1:A10>0,A1:A10,0))
7	5	18.83	Avg of x^2 when x > 0	=SUMPRODUCT(--(A1:A10>0),(A1:A10)^2)/COUNTIF(A1:A10, ">0")
8	3			
9	1			
10	-12			

Chapter 7

In Chapter 12 we learn to use Solver which would tell us that $h = 7.2$" will result in V = 132.2 gals (or 264.4/2).

1. The curve in the chart is made in the normal way from a table with *height* values in one column and *volume* values (in gallons) in another. The worksheet shows that when $h = 0$ the maximum volume is 264.4 gals. Experimentation shows that 7 in. gives a volume of 136.9 gals which is close enough to half the maximum for plotting purposes. The data shown in the following table is added to the chart using the *Copy and Paste Special* technique. The shape is added with *Insert | Shapes*.

x	y
0	136.9
7	136.9
7	0

Of course, the alternative would be to plot $y = \sin(x)+1$ and $y = \cosh(x)$.

2. The two data series on the chart are $y = \sin(x)$ and $y = \cosh(x) - 1$. The first few lines of the table for the chart have these formulas shown below. The trick to making the chart useful is to limit the y-axis maximum to a value of 2.

	A	B	C
3	x	sin(x)	cosh(x)-1
4	=-PI()	=SIN(A4)	=COSH(A4)-1
5	=A4+PI()/5	=SIN(A5)	=COSH(A5)-1
6	=A5+PI()/5	=SIN(A6)	=COSH(A6)-1

Chapter 8

1. In keeping with Hubble's Law, a straight line gives a good fit to this data. The slope (Hubble's constant) is 66.25.

2. (a) Here are two graphical methods. For those who are more comfortable with the idea that "a line of best fit needs to be a *straight* line," the top chart gives $k = 0.0008517$. The more direct approach in the lower chart gives the same result. Note that the intercept was set to zero in the linear chart and to 1 in the exponential chart.

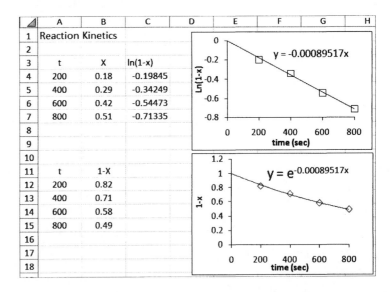

(b) Since we want only the slope we can use LINEST or LOGEST in a single cell and not bother with making them array functions. Note that in this case SLOPE and LINEST are the same when the intercept is not fixed.

Function	k	Formula	Comment
LINEST	8.952E-04	=-LINEST(C4:C7,A4:A7,FALSE)	Force line thru origin
	8.735E-04	=-LINEST(C4:C7,A4:A7,TRUE)	Vary intercept
SLOPE	8.735E-04	=-SLOPE(LN(1-B4:B7),A4:A7)	Vary intercept
LOGEST	8.952E-04	=-LN(LOGEST((1-B4:B7),A4:A7,FALSE))	Force line thru origin
	8.735E-04	=-LN(LOGEST((1-B4:B7),A4:A7,TRUE))	Vary intercept

A full LINEST formula with statistics, or a chart trendline, shows that for a second-order polynomial fit R^2 is 1. So we are justified in using this approach.

3. In the accompanying figure, rows 6 and 7 hold the data to be fitted. In E3:G3 the LINEST function is used to fit the data to $i = at^2 + bt + c$. The differential of this is $i = 2at^2 + b$; the coefficients ($2a$ and b) are stored in F4:G4. The values of di/dt are generated in row 9. The formula in B9 is =$F$4*B6+$G$4. The V values are computed in row 10; in B10 the formula is =$B$3*B9. Row 11 shows the results we obtained earlier with the numerical differentiation formulas. There is reasonable agreement for the interior values but poor agreement for the exterior ones.

	A	B	C	D	E	F	G
1	Numerical Differentiation				Quadratic LINEST Fit		
2				power	2	1	0
3	L	20		f(x)	-20.4853	-0.22087	4.957868
4				f'(x)		-40.9706	-0.22087
5							
6	t	0	0.05	0.1	0.15	0.2	0.25
7	i	4.9550	4.89936	4.73369	4.46172	4.08954	3.62552
8							
9	f'(x)	-0.2209	-2.2694	-4.3179	-6.3665	-8.4150	-10.4635
10	V fit	-4.4173	-45.3879	-86.3585	-127.3291	-168.2996	-209.2702
11	V diff	-22.2560	-44.2620	-87.5280	-128.8300	-167.2400	-185.6080

Chapter 9

A useful test site: http://www.ajdesigner.com/phpdarcyweisbach/pipe_darcy_weisbach_equation_head_loss.php.

2. The following table shows a test worksheet in which the pressure drop is computed both with a worksheet function in column F (using intermediate values in column D) and with a UDF in column G. The code for the UDF is

```
Function DW(length, diam, flow, friction)
    Const g = 32.2
    Pi = 4 * Atn(1)
    vel = 4 * flow / (Pi * diam ^ 2)
    DW = friction * (length / diam) * vel ^ 2 / (2 * g)
End Function
```

	A	B	C	D	E	F	G	H
1	Darcy-Weisbach Equation for water pipes							
2								
3						Darcy-Weisbach		
4	L (ft)	D (ft)	Q (ft³/sec)	V (ft/sec)	f	Formula	Function	test
5	1000	2	200	63.66198	0.02	629.3241	629.3241	TRUE
6	1000	4	200	15.91549	0.02	19.66638	19.66638	TRUE
7	1000	8	200	3.978874	0.02	0.614574	0.614574	TRUE

Since the value of 0.02 is frequently used for water, it might be convenient to be able to omit it in the calling formula whenever that value is to be used.

```
Function DW(length, diam, flow, Optional friction)
    If IsMissing(friction) Then friction = 0.02
    Const g = 32.2
    Pi = 4 * Atn(1)
    vel = flow / (Pi * diam ^ 2)
    DW = friction * (length / diam) * vel ^ 2 / (2 * g)
End Function
```

3. Suitable code is shown here:
```
Function ForceVector(vectorA, vectorB)
  Dim TempVector(2)
  Pi = 4 * Atn(1)
  ForceXA = vectorA(1) * Cos(vectorA(2) * Pi / 180)
  ForceXB = vectorB(1) * Cos(vectorB(2) * Pi / 180)
  ForceX = ForceXA + ForceXB
  ForceYA = vectorA(1) * Sin(vectorA(2) * Pi / 180)
  ForceYB = vectorB(1) * Sin(vectorB(2) * Pi / 180)
  ForceY = ForceYA + ForceYB
  Force = Sqr(ForceX ^ 2 + ForceY ^ 2)
  TempVector(0) = Force
  Theta = Atn(ForceY / ForceX)
  TempVector(1) = Theta * 180 / Pi
  ForceVector = TempVector
End Function
```

4. Suitable code is:
```
Function SciNum(X)
Dim temp(2)
  k = 0
  Do While X > 10
    X = X / 10
```

```
    k = k + 1
     'Debug.Print k
   Loop
   temp(0) = X
   temp(1) = k
    'Debug.Print k
    SciNum = temp
  End Function
```

6. Suitable code is:
```
Function Vmag(myrange)
    For j = 1 To myrange.Count
       Vmag = Vmag + myrange(j) ^ 2
    Next j
    Vmag = Sqr(Vmag)
End Function
```

Chapter 10

2. The following screenshot shows a suitable worksheet. The solution can be found on worksheet *Problem 2* and its associated VBA module the file ProblemsChap10.xlsm on the companion website. Note how a subroutine can return error messages, which is something a UDF cannot do. You could incorporate conditional formatting on the worksheet; row 4 might be red, green, or blue, depending on the value in C4.

⊿	A	B	C	D	E	F
1	Force of drag of fluid over a cylindrically shaped object					
2						
3		Velocity [mph]	Object shape	Length [ft.]	radius [in.]	Fd [lbf]
4		20	concave	12	1	4.6
5						
6						
7			Concave			
8			or			
9			Convex			
10						
11						
12						

3. The following screenshot shows a suitable worksheet. The solution can be found on worksheet *Problem 3* and its associated VBA module the file *ProblemsChap10.xlsm* on the companion website.

	A	B	C	D	E	F	G	H	I	J
1	Soil Contamination Levels									
2			Data				Results			
3		0.021	0.020	0.021	0.021					
4		0.010	0.002	0.010	0.010				Run Macro	
5		0.001	0.001	0.001	0.001					
6		0.005	0.015	0.005	0.005				reset	
7		0.010	0.010	0.010					worksheet	
8		0.020		0.010						
9		0.002		0.010						
10		0.001								
11		0.015								

Chapter 11

1. In the following figure, B11 finds the first approximate value of V using the Ideal Gas Law. Then with a rearranged van der Waals equation we generate successive approximations. The results converge quickly.

	A	B	C	D	E	F
1	Successive Approximations					
2						
3						
4	P	1				
5	T	500		B11	=R_*T/P	
6	a	3.592		B12	=R_*T/(P+a/B11^2)+b	
7	b	0.04267				
8	R	0.082057				
9						
10	Iteration	V (It)	Diff			
11	1	41.0285				
12	2	40.98381	0.044692			
13	3	40.98362	0.00019			
14	4	40.98362	8.11E-07			
15	5	40.98362	3.46E-09			
16	6	40.98362	1.47E-11			
17	7	40.98362	5.68E-14			
18	8	40.98362	0			
19	9	40.98362	0			

$$\left(P + \frac{a}{\bar{V}^2}\right)\left(\bar{V} - b\right) = RT$$

$$\bar{V} = RT\left(P + \frac{a}{\bar{V}^2}\right)^{-1} + b$$

2. There are 11 members in the structure so 11 equations are needed. For nodes 2 through 6 we write two equations—one for the horizontal component of the forces and another for the vertical. We assume each member to be in tension. For node 7 we need only the horizontal equation. The coefficients of the resulting system of equations are shown below. Use matrix math to solve. It is convenient to use named cells for Sin(30), Cos(30), Sin(60) and Cos(60). Only two cells are really needed. The value for f_2 is 28.86 in compression. See the file *Chap11Problems.xlsx* on the companion website for more details.

f_1	f_2	f_3	f_4	f_5	f_6	f_7	f_8	f_9	f_{10}	f_{11}	ext
0.5	0	-0.5	-1	0	0	0	0	0	0	0	-5
-0.866	0	-0.866	0	0	0	0	0	0	0	0	10
0	1	0.5	0	-0.5	-1	0	0	0	0	0	0
0	0	0.866	0	0.866	0	0	0	0	0	0	10
0	0	0	1	0.5	0	-0.5	-1	0	0	0	0
0	0	0	0	-0.866	0	-0.866	0	0	0	0	10
0	0	0	0	0	1	0.5	0	-0.5	-1	0	0
0	0	0	0	0	0	0.866	0	0.866	0	0	10
0	0	0	0	0	0	0	1	0.5	0	-0.5	5
0	0	0	0	0	0	0	0	-0.866	0	-0.866	10
0	0	0	0	0	0	0	0	0	1	0.5	0

Chapter 12

3. This figure shows a worksheet set up for this problem. Solver is used with (i) no target cell, (ii) the *By changing cell* is set to *T* (cell B8), and (iii) the constraint that B9 = B10.

◢	A	B	C	D	E	F
1	Boiling Point of Solvent Mixture					
2						
3		a	b	log(pº)	pº	X
4	Benzene	7.84125	1750	2.0079	101.8396	0.5
5	Toluene	8.08840	1985	1.4717	29.6301	0.3
6	EthylBenzene	8.11404	2129	1.0174	10.4081	0.2
7						
8	Temperature	300 K	27 °C			
9	Total VP	62 torr				
10	Required VP	760 torr				

D4	=B4-C4/T
E4	=10^D4
B9	=SUMPRODUCT(E4:E6,F4:F6)
C8	Named as T

4. The following figure shows a worksheet set up for this

problem. Since this is an optimization problem, each simulation must be solved separately. For the first case, Solver is used with (i) target cell B8 to be minimized, (ii) the *By changing cell* is set to B7, and (iii) no constraint.

	A	B	C	D	E	F
1	Batch Reactor					
2						
3	V	10	20	10	10	10
4	a0	0.1	0.1	0.2	0.1	0.1
5	k	1	1	1	0.5	1
6	tc	0.5	0.5	0.5	0.5	1
7	tr	1	1	1	1	1
8	yield	0.4214137	0.8428274	0.8428274	0.2623129	0.3160603
9						
10	test	0.0837093	0.0837093	0.0837093	-0.119232	-0.098612
11						
12	B8	=(B3*B4*(1-EXP(-B5*B7)))/(B6+B7)				
13	B10	=B7-LN(B7*B5+B6*B5+1)/B5				

5. Let there be two cells named *x* and *y* and a cell with formula =x*y. Cells *x* and *y* are varied to maximize the product subject to the constraint $3y = 18 - 2x^2$ (or that $3y + 2x^2 - 18 = 0$).

6. See the file *SolverCrudeOil.xlsx* on the companion website.

7. The following figure shows a worksheet set up for this problem. Solver is used with (i) no target cell, (ii) the cells to be changed are *a*, *b* and *c* and (iii) the constraint is to make *Objective* = 1.

	A	B	C	D	E	F
1	Puzzle					
2						
3	Digits (variables)			Number		Division
4	a	1		111		37
5	b	1		Sum of Digits		Objective
6	c	1		3		34
7						
8			D4 =B4*100+B5*10+B6			
9			D6 =B4+B5+B6			
10			F4 =D4/D6			
11			F4 =F4-D6			

Chapter 13

3. The answer is 4.27. This may also be found by simple integration. See the workbook *NumericalIntegration.xlsx* on the companion website for a sample worksheet.

4. Using Exercise 4 as a model, you can quickly develop a worksheet and module to get data. With $n = 10$, the result is 0.636294774145023; the error being about 4.13×10^{-7}.

5. The mixed Simpson method gives a value of 2.226 for the area. This data was generated from the function
$$f(x) = \sqrt{2ax - x^2} \text{ with } a = 8$$
The exact result is found to be 2.238 using
$$I = \frac{x-a}{2}\sqrt{2ax - x^2} + \frac{a^2}{2}\cos^{-1}\left(1 - \frac{x}{a}\right)\Bigg]_0^{0.9}$$
The numerical method underestimates by about 0.6%.

8. The trapezoid rule gives 9.79 while integration of the second order polynomial gives 9.75. The agreement is quite good because the curve is very smooth. See the workbook *NumericalIntegration.xlsx* on the companion website.

Chapter 14

2. The following tables show a suitable answer

⊿	A	B	C	D	E	F
1	Euler's Method					
2	Differential equation y'=-2xy with boundary condition y(0)=1					
3	h	0.05				
4						
5	i	x	y	h*f(x,y)	exact	error
6	0	0.00	1.00000	0.00000	1.00000	0.00000
7	1	0.05	1.00000	-0.00500	0.99750	0.00250
8	2	0.10	0.99500	-0.00995	0.99005	0.00495

Row	C	D	E	F
6	0	1	=h*(-2*B6*C6)	=EXP(-(B6^2))
7	=B6+h	=C6+D6	=h*(-2*B7*C7)	=EXP(-(B7^2))

6. A suitable function is:
```
Function MyAmps(E, L, a, b, amp)
  MyAmps = (E / L) - (b / L) * amp ^ 3 - (a / L) * amp
End Function
```
Here are some values so you may check your project before downloading *Chap14Answers.xlsm* from the companion website. The current is changing by about 1E-7 amps when *time* = 0.15 seconds, so we may say this is the steady-state value.

time	amps
0.002	0.128968
0.050	1.164744
0.150	1.179509

7. A suitable worksheet is shown here. Note the excellent agreement with the exact value after nine iterations.

◢	A	B	C	D	E	F	G	H	I
1	Terminal Velocity			Runge-Kutta Method					
2									
3	m	5							
4	g	32							
5	h	1							
6	K	0.125							
7									
8	t	v	k_1	k_2	k_3	k_4	Δv	Exact	% error
9	0	0	32	25.6	27.904	12.534	25.257	0	
10	1	25.2570	16.052	4.3059	13.217	-5.007	7.6820	25.5296	1.07%
11	2	32.9390	4.8756	0.7121	4.286	-2.643	2.0382	33.8322	2.64%
12	3	34.9772	1.4149	0.1652	1.2703	-0.847	0.5731	35.4445	1.32%
13	4	35.5503	0.4044	0.044	0.3653	-0.248	0.1624	35.7213	0.48%
14	5	35.7128	0.1150	0.0122	0.104	-0.071	0.0461	35.7678	0.15%
15	6	35.7588	0.0326	0.0035	0.0296	-0.020	0.0131	35.7755	0.05%
16	7	35.7719	0.0093	0.001	0.0084	-0.006	0.0037	35.7768	0.01%
17	8	35.7756	0.0026	0.0003	0.0024	-0.002	0.0011	35.7770	0.00%

The equation shown in the box:

$$\frac{dv}{dt} = g - \frac{K}{m}v^2$$

Cell	Row 9	Row 10
t	0	=A9+h
v	0	=B9+G9
k_1	=h*(g-(K/m)*B9^2)	=h*(g-(K/m)*B10^2)
k_2	=h*(g-(K/m)*(B9+C9/2)^2)	=h*(g-(K/m)*(B10+C10/2)^2)
k_3	=h*(g-(K/m)*(B9+D9/2)^2)	=h*(g-(K/m)*(B10+D10/2)^2)
k_4	=h*(g-(K/m)*(B9+E9)^2)	=h*(g-(K/m)*(B10+E10)^2)
Δv	=(1/6)*(C9+2*D9+2*E9+F9)	=(1/6)*(C10+2*D10+2*E10+F10)
Exact	=SQRT(m*g/K)*TANH(SQRT(K*g/m)*A9)	=SQRT(m*g/K)*TANH(SQRT(K*g/m)*A10)
% error		=(H10-B10)/H10

Chapter 16

2. This is the data from which Lord Rayleigh inferred the existence of a new gas in air, later called argon. The A series are measurements of nitrogen taken from the air, while the B series are from chemically produced nitrogen. Do they differ significantly? If we use the method of Exercise 7 (see the following table), we, like Rayleigh, come to the conclusion that they do.

For a very detailed statistical treatment see: http://www.econ.vt.edu/Faculty/CVs_&_Research/Aris%20Spanos%20-%20Working%20Papers/Spanos-argon.pdf

◢	A	B	C	D	E	F
1	Comparing repeated measurements					
2						
3		Data			Calculations	
4	A	B			A	B
5	2.31017	2.30143		mean	2.3101	2.2995
6	2.30986	2.29890		st dev	0.0001	0.0014
7	2.31010	2.29816		ssd	1.221E-07	1.332E-05
8	2.31001	2.30182		n	7	8
9	2.31024	2.29869		s pooled	0.0010	
10	2.31010	2.29940		t expt	20.214	
11	2.31028	2.29849		prob	95%	
12		2.29889		df	13	
13				t theory	2.160	
14				outcome	Reject Null hypothesis	
15						
16					p method	
17				p expt	0.000	
18				t-test	0.000	
19				alpha	0.05	
20				outcome	Reject Null hypothesis	

3. The tires were placed one A and one B on each car, so a paired test is appropriate. The following figure shows the results.

D	E	F	G	H	I	J	K
	t-Test: Paired Two Sample for Means						
		X	Y				
	Mean	10.24	9.76				
	Variance	1.733	1.763				
	Observations	5	5				
	Pearson Correlation	0.998034					
	Hypothesized Mean Difference	0					
	df	4					
	t Stat	12.82854					
	P(T<=t) one-tail	0.000106					
	t Critical one-tail	2.131847					
	P(T<=t) two-tail	0.000213					
	t Critical two-tail	2.776445					

t-stat > t-critical one-tail We may reject the null hypothesis. At 95% confidence level, A tires have a mean wear that it greater than B's

4. From the following figure we see that about 97% are expected to have a lifetime of 750 hours or less. Maybe the manufacture should rethink the warranty! Perhaps he meant 450? Had he Googled with *Normal Distribution Calculator* he would have known better!

◢	A	B	C	D	E	F	G
1	Widget Lifetimes						
2							
3	Mean	585		Hours	cumulative p	number	
4	Stdev	89		750	96.81%	1,936,251	Difference
5				600	56.69%	1,133,841	669,849
6				700	90.18%	1,803,690	
7							
8				E4 = NORMDIST(D4 , Mean , Stdev , TRUE)			
9				E5 = NORMDIST(D5 , Mean , Stdev , TRUE)			
10				E6 = NORMDIST(D6 , Mean , Stdev , TRUE)			
11				F4 = E4 · 2000000			
12				F5 = E5 · 2000000			
13							
14				F6 = E6 · 2000000			
15				G5 = F6 - F5			

Index